Biomedical Ethics Reviews • 1985

Biomedical Ethics Reviews

Editors

James M. Humber and **Robert F. Almeder**

Biomedical Ethics Reviews • 1985

Edited by

JAMES M. HUMBER and ROBERT F. ALMEDER

Humana Press • Clifton, New Jersey

Library of Congress Cataloging in Publication Date

Main entry under title:

The Library of Congress has cataloged this serial publication as follows:

Biomedical ethics reviews.—1983-   —Clifton, N.J. :
Humana Press, c1982-
v.; 25 cm.—(Contemporary issues in biomedicine, ethics, and society)
Annual.
Editors: James M. Humber and Robert F. Almeder.
ISSN 0742-1796 = Biomedical ethics reviews.

1. Medical ethics—Periodicals. I. Humber, James M.
II. Almeder, Robert F. III. Series.
[DNLM: 1. Ethics, Medical—periodicals. W1 B615 (P)]
R724.B493 174'.2'05—del9 84-640015
AACR 2 MARC-S

Crescent Manor
PO Box 2148
Clifton, NJ 07015

Printed in the United States of America

Preface

Biomedical Ethics Reviews: 1985 is the third volume in a series of texts designed to review and update the literature on issues of central importance in bioethics today. Four topics are discussed in the present volume: (1) Should citizens of the United States be permitted to buy, sell, and broker human organs? (2) Should sex preselection be legally proscribed? (3) What decision-making procedure should medical personnel employ in those cases where there is a high degree of uncertainty? (4) What do we mean when we use the terms "health" and "disease"? Each topic constitutes a separate section in our text; introductory essays briefly summarize the contents of each section.

Bioethics is, by its nature, interdisciplinary in character. Recognizing this fact, the authors represented in the present volume have made every effort to minimize the use of technical jargon. At the same time, we believe the purpose of providing a review of the recent literature, as well as of advancing bioethical discussion, is admirably served by the pieces collected herein. We look forward to the next volume in our series, and very much hope the reader will also.

James M. Humber
Robert F. Almeder

Editors' Note

All of the articles published in the first four volumes of *Biomedical Ethics Reviews* were written upon request of the editors. We now plan to change this editorial policy. Although we may request individuals to write articles on certain topics, the vast majority of essays published in *Biomedical Ethics Reviews* will henceforth be selected from a pool of submitted articles. Each year we will specify discussion topics for the succeeding year's issue of *Biomedical Ethics Reviews* and put out a general call for papers. Manuscripts will be assessed for quality and chosen for publication by the editors and members of *Biomedical Ethics Reviews'* Editorial Board. *Biomedical Ethics Reviews: 1987* will be the first volume in our series of texts to publish articles chosen through open submission. Topics to be discussed in this volume include:

1. Animals as a source of human transplant organs.
2. The nurse's role: rights and responsibilities.
3. Should human research be conducted at private, profit-making hospitals?
4. Prescribing drugs for the aged and dying in the health care setting.

Biomedical Ethics Reviews is interdisciplinary in character, and scholars from all disciplines are encouraged to submit manuscripts for publication in *Biomedical Ethics Reviews: 1987*. All contributions should be between 2500 and 10,000 words double spaced, submitted in duplicate, and accompanied by a 250–500 word abstract. Authors must follow the style established in the published volumes, and special attention should be given to references and footnotes, which should be collected at the end of the essay. Manuscipts intended for publication in *Biomedical Ethics Reviews: 1987* must be received by the editors no later than June 31, 1986. All communications should be addressed to: James Humber and Robert Almeder, Department of Philosophy, Georgia State University, University Plaza, Atlanta, Georgia 30303-3083. Authors whose essays are published in *Biomedical Ethics Reviews* will receive royalties according to terms specified by Humana Press, Inc., Clifton, NJ.

Contributors

WILLIAM BECHTEL • *Department of Philosophy, Georgia State University, University Plaza, Atlanta, Georgia*

MARTIN BENJAMIN • *Department of Philosophy, Michigan State University, East Lansing, Michigan*

BERNARD BOXILL • *Department of Philosophy, University of South Florida, Tampa, Florida*

SAMUEL GOROVITZ • *Department of Philosophy, University of Maryland, College Park, Maryland*

HELEN BEQUAERT HOLMES • *Science and Society Program, Biology Centre, Heren, The Netherlands*

JAMES HUMBER • *Department of Philosophy, Georgia State University, University Plaza, Atlanta, Georgia.*

SUZANNE POIRIER • *Humanistic Studies Program, The University of Illinois at Chicago, Chicago, Illinois*

MARY ANNE WARREN • *Department of Philosophy, San Francisco State University, San Francisco, Calfornia*

Contents

Medical Decisionmaking Under Uncertainty

Concepts of Health and Disease

Section I

Buying, Selling, and Brokering Human Organs

Introduction

Professor Samuel Gorovitz's testimony before the House Subcommittee on Investigations and Oversight serves as the lead article in this section. In his testimony, Gorovitz argues that the buying and selling of human body parts should be legislatively proscribed. Gorovitz's attack is two-pronged. On the one hand, he opposes a specific plan by H. Barry Jacobs to broker human kidneys; on the other, he argues that any enterprise similar to that proposed by Jacobs ought to be prohibited. As against Jacobs' plan, Gorovitz argues: (1) That risks to organ donors are greater than Jacobs admits, and (2) that the scheme not only makes a mockery of informed consent, but is also plagued by quality control problems. When he argues against the establishment of any commercial market in human organs, Gorovitz contends: (a) That a commercial market in human organs would exploit the underprivileged both at home and abroad and damage our international credibility, and (b) that such a market would perpetuate the injustice of allocating vital medical resources according to the ability to pay.

In his article, "Coercion, Paternalism, and the Buying and Selling of Human Organs," Professor James Humber critically evaluates Gorovitz's arguments and concludes that none is sound. Furthermore, Humber claims that Gorovitz's position denigrates the poor and invokes a dangerous governmental paternalism. In the end, Humber argues that if the poor and underprivileged in our society are to be viewed as autonomous, rational beings, capable of membership in the moral community, it is they, and not the government, who should decide whether or not their organs will be sold.

In "Global Objections to Kidney Sales," Gorovitz responds to Humber's criticisms. First, Gorovitz admits that the arguments offered in his testimony before the House Subcommittee, when considered independently, are inconclusive. However, Gorovitz insists that when those arguments are taken together, they create a preponderance of evidence against the establishment of a commercial market in human organs. Second, Gorovitz examines Humber's criticisms individually, and rejects each in turn. Finally, Gorovitz argues that Humber misunderstands his position. Gorovitz claims that his testimony before the House Subcommittee was not designed to demonstrate that there is something wrong with the buyer–seller relationship in particular instances of organ sale, but rather to show that there is something wrong with a society in which such an exchange can make sense.

Testimony

House Subcommittee on Investigations and Oversight

Samuel Gorovitz

Congressman Gore and Members of the Subcommittee, I welcome the opportunity to discuss with you some of the problems associated with recent developments in transplant surgery. I take it that these facts, at least, are uncontroversial: that there has been a sharp, recent increase in our capacity to transfer living tissues of various kinds successfully from donors to recipients, that there is in consequence a large and growing shortage of transplantable materials, and that controversial economic, political, and moral issues swirl around our efforts to respond to this new situation.

How can we best meet the vital needs of patients who require transplant surgery, while respecting the various related interests and concerns which come into play? We are faced here with choices which in the words of Richard Titmus, "lead us, if we are to understand these transactions in the context of any society, to the fundamentals of social and economic life." The question of how to close the gap between the demand for and the supply of transplantable organs is no less than the question of what sort of society we wish to advocate, endorse, and nurture.

The range of possible responses is great. We have heard proposals to presume consent by prospective donors in the absence of clear evidence to the contrary, proposals to establish commercial markets in organs, and proposals to increase the efficiency of present approaches through devices ranging from tax incentives to public education. I am not aware that anyone has yet proposed seriously that transplantable organs should be made available without regard to the wishes of the person whose organs are at issue—but such a proposal would not surprise me.

The focus of this testimony is the specific question: What should the government do, promote, permit, or prohibit in respect to organ transplantation? Any response to that question must rely on a broader conception of the proper role of government generally. So I want to be clear at the outset about some of my convictions in that regard.

It is not the responsibility of government to be the solution of first resort to the problems of contemporary society; rather, the private sector is our best hope for meeting a broad range of needs. The government has a responsibility to step in only where it must, to safeguard the public interest. Further, the government should exercise great caution in enacting prohibitions on behavior. Only where it can sustain a persuasive justification may it properly constrain the behavior of citizens; it has no business prohibiting actions merely because they are offensive to the sensibilities of a portion of the citizenry, or because they could conceivably lead to more serious abuses in the future. Nor may it require actions simply because they would be in the public interest. Requiring actions (such as the payment of taxes or participation in national defense in wartime) or prohibiting actions (such as violation of the civil liberties of citizens) requires strong justification indeed.

It is for this reason that I argued before you last year that the appropriate role for government in respect to the shortage of organs is catalytic rather than coercive. To require the donation of cadaver organs would be to ride roughshod over the rights of individuals to exercise discretion over the disposition of their bodily parts. Even to presume consent in the absence of dissent would be to place the burden where it does not belong. Those who prefer not to donate organs, for reasons of religion, superstition, or squeamishness, or for no reason at all, would be cast into a defensive position in which they might feel hard pressed to protect themselves and their families against intrusions of a most intimate sort.

Yet the problem remains and grows, so something must be done. An ideal solution would be a massive shift in national sentiment about transplantation—a shift that would greatly increase participation in donation plans and would also greatly diminish the barriers, psychological and economic, to participation by the medical profession in efficient collection and distribution of organs. Ideal solutions are always elusive, and it is prudent to ask whether we can even approximate to them in fact. With respect to organ transplantation, we do not yet know the answer to that question. The large shortage of organs is too recent, and our current modes of response to that shortage are too unsystematic, for us to know what we can yet achieve in the way of an enlightened collective response.

It is time now to put that question to the test. The newly created American Council on Transplantation may become an effective instrument for rationalizing our methods of collecting and distributing organs and of increasing public participation in donation plans—but that will require it to have significant financial and institutional backing, a firm and energetic resolve to meet its objectives, and a fair bit of good luck. Its prospects of success will be greatly enhanced by the passage of HR4080, which, without being coercive or intrusive, fosters a major increase in our structural capacity to achieve an adequate solution based on a voluntary and altruistic response to the plight of potential transplant recipients.

Such an approach must be given every reasonable chance of success, for the alternatives are grim. The demand for transplantation will continue to increase, as will the variety of transplantable tissues. Today, we focus mainly on kidneys, corneas, and livers, knowing that lungs and hearts are also transplantable. But skin, bone and muscle are transplantable, too, and recent successes in the reattachment of digits and limbs foreshadow the transplantation of such parts in response to major trauma. It would be naive not to realize that we are at the beginning of the problems associated with our newly developed capacities of medical and surgical intervention.

And what are those grim alternatives? One of the worst would be a governmental takeover of the whole domain, responding to national shortages with national systems of allocation in accordance with national criteria, supported by mandatory and intrusive processes of collection. The disadvantages of such a scheme, I trust, need no elaboration here. However, a comparable peril exists on the other side. For, another alternative to the present shortage is to allow a commercial market to flourish, linking supply and demand through the mechanisms of free enterprise. And the disadvantages of that scheme do require some elaboration.

Already, H. Barry Jacobs of Virginia has established a business for the commercial brokering of kidneys. As you doubtless know, he has proposed to commission the sale of kidneys from persons, initially in the Third World, for whatever price is needed to induce them to sell, and then to broker the kidneys to affluent Americans who need transplantable organs and can meet the costs. The brokerage fees will make the enterprise, in Jacobs' own words, "a very lucrative business."

It is important to think carefully about this plan, more deeply than at a level of initial reactions. For it raises many questions that go beyond the immediate need for more kidneys. A frequent initial response to the Jacobs scheme is that it is morally repugnant. But so might an appendectomy be to one who is innocent of the benefits of abdominal surgery. Is the distaste engendered by Jacobs' scheme not also a result of a shallow reaction, prompted by an unfamiliar solution to a new, but vital problem? That, at least, is what he would have us believe; he defends his scheme with a veneer of humanitarianism, public service, and the American way. Before we legislate him out of business, we had best consider the merits of his case. For it is not simply a question of the economic ambitions of one ex-practitioner from Virginia, it is a question of determining important features of the distribution of vital resources for the challenging years to come.

People beset by extreme poverty, malnutrition, and ignorance live in desperate circumstances. So, too, do those with end-stage renal failure. Surely, one aspect of the Jacobs scheme is that it is profiteering on the desperation of these two groups. But if, as he claims, all related transactions are to proceed by the voluntary actions of fully informed and uncoerced adults, we must pause before concluding that there is a legitimate public interest in prohibiting such exchanges. At least, as Jacobs points out, his plan will deliver kidneys to people who need them, and cash to people who need it, possibly to their mutual benefit. And with the traditionally admirable flexibility of the free-market system, the scheme can function long before the catalytic efforts of the government or the American Council on Transplantation can take effect.

Many assumptions in the Jacobs scheme are open to challenge. The risks to donors are greater than he has admitted, for example. And the scheme make a mockery of informed consent, as is evident to anyone familiar with Federal regulations protecting human research subjects, which reflect a sensitive awareness that desperate circumstances can be implicitly coercive, and that the provision of excessive inducements to the oppressed can constitute a violation of their autonomy. And there are problems of quality control that might be insuperable.

But we miss the most fundamentally important issues if we focus our attention on such weaknesses in the proposal that there be a commercial market in kidneys. Very much larger matters are at stake.

There are various standards for judging the greatness of a society. One measure is by the peaks of its achievements in the arts and culture or in technology. Another measure is the average material standard of living of its people. A third is the scope of its territorial authority, and so on.

I have always thought that one appropriate standard for judging the greatness of a society is that of how it treats those whom it treats least well. The analog at the level of the family is compelling, at least. No matter how we admired the talented, affluent, accomplished family next door, our judgment of them would plummet if we discovered that they had one family member whom they abused, whose interests they ignored, whose needs left them unmoved, and whom they exploited to their own maximum advantage. Such a discovery would provide us with important information about their character and integrity—about their sense of connectedness to one another and their sense of justice within a social structure. Judged by the analogous criterion, American society does not yet live up to its loftiest ideals.

Another criterion for judging the greatness of a society is the way it treats its most seriously disadvantaged. (This criterion is related to, but is not the same as, the previous one.) People in grinding poverty and those beset by life-threatening illness are surely in highly disadvantaged circumstances. What societal response to their plight do we wish to endorse?

A free-market model is based on the values of competition, individual initiative, and the elasticity of supply and demand in response to market forces. But medical need is no respector of success in the world of commerce. The poor are more likely, not less likely, to be seriously ill, and their ability to obtain medical care is seriously compromised by their poverty. To distribute vital resources according to ability to pay is to set aside all concern for medical need as the primary determinant of access. It is to set aside considerations of compassion and cooperation and abandon the effort to fashion a society in which mutual supportiveness is our response to desperation. It is to sanction the expansion of unfettered commercialism into dimensions of life which could just possibly provide us the opportunity to achieve a greater sense of community and of national purpose than we have previously known, except in the face of external threat. It is to ask far too little of ourselves.

The argument for a commercial market in kidneys might have greater force had we put ourselves to the test, and failed. But we are just now acknowledging a new national need, and HR4080 seeks to fashion a constructive response to that need. It is far too soon to judge that response a failure; it is too soon even to decide, as we may be able to decide a year from now, whether that response has been able to match the sizzling pace of new medical developments.

The only adequate barrier to the commercialization of life, as proposed by Jacobs, is a new legislative prohibition. I support that prohibition for many reasons, and I do so as one reluctant to endorse any unnecessary restriction on individual liberty. Such prohibition, however, is necessary to put to the test our capacity as a nation to meet the present shortage, and to fashion ways to deal with future shortages, with a due respect for the dictates of equality and social justice. The credo of the French revolution—liberty, equality, and fraternity—can remind us at times like this that we are well advised to temper our passionate and worthy defense of liberty with a due consideration of the social context without which our liberty would be a tragically empty achievement.

An additional reason for supporting the prohibition derives from the symbolic significance of the proposed market in organs. At a time when we urgently need to nurture good relations with the nations of the Third World, our international credibility would be dealt a severe blow by our tolerance of a plan according to which the poor in underdeveloped countries were exploited as a source of spare parts for rich Americans. Our antagonists behind the iron curtain would love such a public relations windfall—and they would be right.

In the third world, it is unlikely that strong restrictive action will effectively prevent the plundering of peasants' parts for profit. Their public health concerns are still centered on problems of sanitation, nutrition, and infectious disease. If there are to be effective controls, they must be at our end. But it is not surprise, nor inappropriate, that the world's most highly developed nation should bear the burdens of exercising responsibility over medical science's most advanced capacities.

Nor should we be swayed by the claim that commercial markets in live organs will develop elsewhere no matter what we do, and that wealthy Americans will be the recipients in any case. Whether or not that is so remains to be seen, but has no bearing on the fact that we must act rightly as we can best judge the right. If we want the world to be inspired and informed by our example as humane and just society, we must be prepared to provide that example.

I am not concerned merely with the prospects for international exploitation and the damage that threatens for our image abroad. I have pressed Barry Jacobs in debate to explain why he prefers to seek organs elsewhere, rather than from among America's downtrodden—the street people in New York, the unemployed migratory farm laborer, the inner city destitute, the most impoverished of our native Americans on reservations. And he has replied that he will go abroad for kidneys only if these American sources are inadequate to meet demand. Thus, would he have us turn on our own poor to seek relief for our well-to-do.

I am concerned, of course, with what such markets would do to those whose destitution and desperation might move them to sell bodily parts in the hope of gaining a foothold for the climb out of poverty. But I am concerned even more about what such behavior would do to the rest of us, and what it would reveal about our compassion, our commitment to equality, our willingness to face common problems with collective resolve, our capacity to make voluntary efforts in the public interest, and more.

That the poor are exploited is unarguable. That their poverty seems intractible is a continuing tragedy of our unprecedentedly affluent society. I hope that history will be able to judge us as a society that never abandoned its struggle to eliminate that poverty, that strove ever to enhance and enrich its respect both for individuals and for their capacity for mutual aid, and that faced the problems of an awesome new technology with humanity and efficiency both, rather than as merely another commercial opportunity. I believe that there is a legitimate public interest in striving to bring this about.

Mr. Gore. Thank you. That was well worth waiting for. I appreciate such a powerful and eloquent statement.

I take it that there is unanimity on this panel that buying and selling human organs would be a bad idea, on balance?

Dr. Brenner, the implications of the statement you presented I think obviously bear further scrutiny. We cannot do that in detail here, but your conclusion is that, at the present time, your findings are not sufficient to lead you to advise against living-related donations, but they do lead you to believe there should be greater care, and certainly more scrutiny, in order to determine whether these results are, as you put it, the tip of an iceberg or not.

Certainly they challenge any assumption that the risks of organ donation by a living donor are far from negligible. They are somewhere between negligible and prohibitory, but they may be more significant than we have assumed in the past. Is that a fair summary?

Dr. Brenner. very fair.

Mr. Gore. . . . I think that your statements are all very eloquent, and I appreciate the obvious time and effort that you have put into them.

I, too, believe that there are larger questions involved here. We do have to decide what kind of society we are going to become. We are experiencing a pace of change unprecedented in all of human history. The recent advances in organ transplantation serve as only one small example of how advances in science and technology are challenging the ethical and legal underpinnings of our civilization.

If we are going to adapt well, we're going to have to engage in a lot of discussions like the one we have had here today, and we are going to have to make some hard choices, like the ones posed by Dr. Jacobs' plan.

I think that HR4080 makes the correct choice with regard to the buying and selling of human organs. I am comforted by the fact that a panel as distinguished as this one supports that judgment.

I do not believe it is a completely open-and-shut case, and I understand that you all appreciate the subtleties involved in these ethical judgments. But I think, on balance, the answer is a fairly decisive one. Again, I am comforted by the fact that you agree.

Let me thank you all again for your testimony here and for your assistance in our exploration of this issue. We are going to continue as a subcommittee to try to improve the way our country deals with organ transplantation, and your contribution today has been extremely significant.

Thank you very much.

Coercion, Paternalism, and the Buying and Selling of Human Organs

James Humber

Introduction

Professor Samuel Gorovitz contends that Congress should pass legislation prohibiting the commercial brokering of human organs.[1] The fact that he holds such a view is significant, for Gorovitz believes "the government should exercise great caution in enacting prohibitions on behavior," and that restrictive legislation ought to be enacted only when the government "can sustain a persuasive justification" for doing so.[1] It is only because Gorovitz believes there is a persuasive justification for prohibiting the buying and selling of human body parts that he advocates proscriptive legislation, and in his testimony before the House Subcommittee on Investigations and Oversight he goes to some pains to develop that justification. Like Gorovitz, I believe our government needs a compelling justification if it is to properly impose legislative restrictions upon the behavior of its citizens. On the other hand, I do not believe that Gorovitz's testimony before the House Subcommittee provides that justification; indeed, I believe there are good reasons that a commercial market in human body parts should not be legislated out of existence. In what follows, I am going to support both of these views. I shall begin by critically evaluating Gorovitz's arguments against the establishment of a commercial market in human body parts. Once the deficiencies in these arguments have been noted, I will explain why I think it would be wrong to prohibit the buying and selling of human organs.

Gorovitz's Arguments

Gorovitz uses two types of argument to show that it would be wrong to permit human organs to be bought and sold. The first type of argument is directed against specific aspects of a plan proposed by H. Barry Jacobs to broker human kidneys. Gorovitz describes Jacobs' plan as follows:

> [Jacobs] has proposed to commission the sale of kidneys from persons, initially in the Third World [and eventually from the poor in our society], for whatever price is needed to induce them to sell, and then to broker the kidneys to affluent Americans who need transplantable organs and can meet the costs. The brokerage fees will make the enterprise, in Jacobs' own words, "a very lucrative business."[1]

In opposition to this scheme, Gorovitz argues (1) that there are problems of quality control that might prove insuperable, (2) that the risks to donors are greater than Jacobs admits, and (3) that "the scheme makes a mockery of informed consent, as is evident to anyone familiar with federal regulations protecting human research subjects, which reflect a sensitive awareness that desperate circumstances can be implicitly coercive, and that the provision of excessive inducements to the oppressed can constitute a violation of their autonomy."[1]

Gorovitz does not simply want Jacobs' kidney brokerage firm to be legislated out of existence. Rather, he wants Congress to pass legislation prohibiting *any* enterprise similar to that proposed by Jacobs, and it is in support of this position, I think, that Gorovitz offers his second set of arguments. Unfortunately, the arguments in this set are not as clear or well delineated as those included in his first group; in the end, however, Gorovitz seems to be using the following arguments to support his view.

First, Gorovitz offers two standards for judging the greatness of a society: (1) a society can be judged by how well it treats those whom it treats least well, and (2) a society can be judged by how it treats its most seriously disadvantaged. Gorovitz does not seem to feel that contemporary American society fares well when judged by either of these standards. He uses the analogy of an affluent family that mistreats one of its members to argue that our society (to some degree, at least) exploits, abuses, and ignores the interests of those whom it treats least well.[1] He also claims that distributing vital medical resources according to the ability to pay effectively denies poor people access to proper medical care and "sets aside considerations of compassion and cooperation, and abandons the effort to fashion a society in which mutual supportiveness is our response to desperation." Since our society presently allocates medical resources principally on the basis of one's ability to pay, the clear implication is that our society cares little about its most seriously disadvantaged, and hence falls short when judged by standard (2).

Although America cannot be said to be "great" when judged by standards (1) and (2), current demand for human transplant organs gives our nation an opportunity to improve its character. If we allow commercial markets in human organs to flourish, we continue to exploit and abuse those whom we treat least well, and persist in setting aside considerations of compassion and cooperation when dealing with the destitute. On the other hand, if we prohibit the buying and selling of human body parts, we refuse to continue to act in these ways. In addition, we give ourselves an opportunity to "test our capacity as a nation to meet the present shortage, and fashion ways to deal with future shortages, with due respect for the dictates of equity and social justice."[1,2] Since prohibition of commercial markets in human organs is necessary to put ourselves to the test and improve our character as a nation, it is obvious that we should enact proscriptive legislation. Indeed, to do otherwise is, in Gorovitz's words, "to ask too little of ourselves."

Gorovitz is opposed to the establishment of brokerage firms in human organs, not only because brokers in these firms would seek to secure body parts from the poor and underprivileged in our society, but also because they would attempt to obtain organs from person in the Third World. If it would be wrong to exploit our country's destitute, it would also be wrong to exploit the poor in other lands. In addition, Gorovitz believes there is a more practical reason that such action ought not be permitted. As Gorovitz puts it " . . . our international credibility would be dealt a severe blow by our tolerance of a plan according to which the poor in underdeveloped countries were exploited as a source of spare parts for rich Americans. Our antagonists behind the iron curtain would love such a public relations windfall. . . . "[1]

Gorovitz's Arguments: A Critical Analysis

Gorovitz argues that Jacobs' kidney brokerage firm ought to be legislated out of existence because, among other things, it could give rise to insuperable problems of quality control. There are, however, at least two problems with this argument.

First, Gorovitz offers no evidence whatsoever for the view that Jacobs' scheme could produce "insuperable" quality control problems. At best, then, Gorovitz has done nothing more than note that there is a *possibility* of such problems arising. If this is so, however, Gorovitz's argument does not show that Jacobs' ought to be prohibited from conducting business, but only that his firm should be closely monitored to ensure that insuperable quality control problems do not arise.

Second, contrary to what Gorovitz asserts, there seems to be good reason to suppose that Jacobs' kidney brokerage firm will not be plagued by insurmountable quality control problems. At present, family planning clinics operate in virtually all large cities in the United States, and many of these clinics supervise second trimester abortions. Second trimester abortions are not danger-free procedures, but as far as I know, no "insuperable" problems of quality control have risen at these clinics. If proper quality control can be guaranteed in these cases, it is difficult to see why it could not be guaranteed at a transplant clinic. Indeed, any broker of human kidneys who did not do his or her utmost to ensure proper quality control would be a very poor businessperson, for if a high percentage of kidney donors died, or were irreparably harmed when donating their kidneys, brokers would find it difficult to convince others to sell their organs. In short, the profit motive should work to ensure a high degree of quality control in transplant clinics—perhaps even higher than that presently exhibited in family planning clinics.

A second reason Gorovitz opposes Jacobs' scheme for buying and selling kidneys is that he believes risks to donors are greater than Jacobs admits. Once again Gorovitz offers no evidence for his view, but let us ignore this point and assume, for the sake of argument, that Gorovitz is correct. In this case, we can interpret Gorovitz's argument in either of two ways: (1) Gorovitz might be arguing that Jacobs' brokerage firm ought to be legislated out of existence because members of that firm will fraudulently obtain kidneys by misinforming kidney donors concerning the degree of risk, or (2) Gorovitz might be claiming that the risks are so high that the government is justified in prohibiting its citizens from taking those risks.

Let us examine each possibility in turn. If (1) correctly represents Gorovitz's view, Gorovitz's argument does not show that Jacobs' firm ought to be legislated out of existence, but rather that safeguards need to be implemented to ensure that representatives of Jacobs' firm do not misrepresent risk when dealing with prospective donors. Gorovitz admits that unnecessary restrictions on individual liberty are undesirable.[1] Thus, unless Gorovitz can show that legislating Jacobs' business out of existence is necessary to protect prospective organ donors from fraud, he will not have a sound justification for enacting such legislation. It seems clear, however, that Jacobs' customers can be protected from misrepresentation without Jacobs' corporation having to be dissolved. Let us say, for example, that Jacobs believes risks to organ donors is of magnitude X. The government disagrees, and assigns a much higher risk factor. In this case the government could demand that members of Jacobs' firm use the government figures when informing potential kidney donors of risk, and require that a videotape record be kept of all communications between representatives of Jacobs' firm and prospective donors. This would protect Jacobs' customers against any possible misrepresentation of risk, and do so without legislating Jacobs' firm out of existence.[3]

When Gorovitz states that risks to donors are greater than Jacobs admits, he might be accepting argument (2) rather than argument (1). If this is the position Gorovitz means to defend, however, he is arguing for a most vicious form of government paternalism. It is not the government's job to determine the degree of risk its citizens are permitted to sustain in their lives. If I want to jump the Grand Canyon in a jet powered motorcycle for the sake of monetary gain, I should be permitted to do so, provided I am rational, understand the risks involved, and am not coerced in my decision. And this, I think, takes us to the very heart of the matter, for I do not believe Gorovitz wants to argue that the government has a right to prohibit dangerous activities. Rather, his main concern seems to be that if Jacobs' firm is allowed to do business, the poor will be coerced into selling their organs. It is to this aspect of Gorovitz's argument, then, that we must now turn.

Gorovitz's final argument against Jacobs' business venture is that the scheme "makes a mockery of informed consent." In support of this claim, Gorovitz appeals to the Federal regulations protecting human research subjects he tells us that these guidelines exhibit "a sensitive awareness that desperate circumstances can be implicitly coercive," and that they recognize that "excessive inducements to the oppressed can constitute a violation of their autonomy." What Gorovitz does not do, however, is explain why we ought to extend the Federal regulations to cover Jacobs' kidney donors. The Federal regulations govern research activities and apply to research subjects. Jacobs' enterprise is not research oriented, and by no stretch of the imagination could his firm's organ donors be called research subjects.

Even if the Federal regulations protecting human research subjects are not extended to cover Jacobs' kidney donors, Gorovitz's basic challenge to Jacobs' scheme remains. That is, does Jacobs' business venture make a mockery of informed consent? If informed consent to a certain course of action is to be assured, at least two conditions must be satisfied: (a) the person consenting must understand the risks involved in the course of action being contemplated, and (b) the person consenting must not have his or her decision coerced or manipulated. It is quite likely that Gorovitz is worried about satisfaction of condition (a) when he claims that Jacobs underestimates the risks of organ donation; however, we have seen that safeguards can be implemented to ensure that potential kidney donors are correctly informed concerning the risks involved in organ donation.[4] This leaves condition (b), and on this issue Gorovitz clearly is of the opinion that the poor are in such desperate circumstances that their decision to sell a body part could be coerced, manipulated, or controlled by excessive inducements from representatives of Jacobs' firm. That this is not the case, however, can be shown both by an understanding of our ordinary use of "coercion," and by a detailed analysis of the concept.

First, consider the following cases. Person P is an out-of-work ghetto dweller who agrees to sell his kidney for $100; person PP lives in conditions exactly similar to those of P, but agrees to sell a kidney for $100,000. Now if we accept Gorovitz's understanding of "coercion," we must say that it is more likely that PP was coerced into selling his or her kidney than P was, for PP was subjected to a greater financial inducement than P. But would anyone ordinarily make such a judgment? I think not. Indeed, I believe most people would say that it is more probable that P was coerced into selling a kidney than that PP was manipulated. After all, PP's actions seem rationally justifiable, whereas P's actions appear, if not irrational, at least mystifying. Surely, then, if anyone was manipulated in the above business transactions, it was P rather than PP.

If an appeal to our ordinary use of "coercion" casts doubt on Gorovitz's position, a more detailed analysis of the concept deals his view a death blow. When, after all, does it make sense to say that one person controls, manipulates, or coerces another's actions? One of the most plausible analyses of the conditions under which we apply these terms is as follows:

A person C controls the behavior of another person P *if and only if:*

(1) C intends P to act in a certain way, A;
(2) C's intention is causally effective in bringing about A; and
(3) C intends to ensure that all of the necessary conditions of A are satisfied.[5]

If we accept this analysis, we can see why it is not correct to say that Jacobs' kidney brokers would coerce or manipulate the poor into selling their organs. Let us say, for example, that kidney broker (K) seeks to buy the kidney of a poor person (P). K may well intend that P sell his or her kidney, and K's intention may even be causally effective in bringing about that sale. Still, K does not intend to ensure that *all* necessary conditions for sale are satisfied. To be sure, K approaches P because K has money to exchange for a kidney, and K knows P is more likely than a rich person to desire added income. But K does not create P's poverty or the desire for added income that is generated by that poverty; these conditions are merely "givens" that K encounters in his transactions with P. As such, it is not correct to say that K coerces, controls, or manipulates P's decision to sell a kidney. Indeed, to hold otherwise one would have to claim that business transactions are coerced *whenever* a salesperson makes a sale by appealing to a customer's in dependently generated needs and desires, and this is absurd.

We have examined Gorovitz's first set of arguments against Jacobs' business scheme and seen that none succeeds in its purpose. It is now time to turn our attention to Gorovitz's second set of objections, and see whether any of these arguments show that it would be proper to prohibit organ brokerage firms from doing business.

One reason Gorovitz opposes any scheme for buying and selling human organs is that he feels businesses engaged in such an activity would exploit the downtrodden. When we say that a person has been exploited, we ordinarily mean that he or she has been taken advantage of, or has been used as a means for someone else's ends. Now if representatives of organ brokerage firms misrepresented risk to their customers, or if they coerced the decisions of prospective donors, a good case could be made that these firms were "using" organ donors for their own or others' (namely, buyers') selfish ends. We have seen, however, that safeguards can be instituted to ensure that brokerage firms do not misrepresent risk, and that there is no reason to assume that brokers will coerce customers into selling their body parts. On the other hand, when Gorovitz claims that the poor will be exploited by organ brokers, he might mean that these individuals will take advantage of the poor by inducing them to sell their organs for less than what they are really worth. But if this is what Gorovitz means, how does one compute the "real worth" of a body organ? Without a computation procedure for determining the true worth of any given person's body parts, it is difficult to see how one could argue that organ brokerage firms will exploit the poor by getting them to sell their organs for less than full value. As far as I know, no such computation procedure has been developed. Indeed, even if a procedure were formulated so that the true value of any human's parts could be ascertained, one could not know, *a priori,* that organ transplant firms would induce the poor to sell their organs for less than full value. To know this, one would have to examine sales data from organ brokerage firms, and this requires that enterprises of this sort exist and do business. At best, then, the version of the "exploitation" argument that we are now considering shows only that the underprivileged in our society *could* be exploited by organ transplant firms, and Gorovitz himself admits that "[the government] has no business prohibiting actions merely because . . . they could conceivably lead to . . . serious abuses in the future."[1]

One possibility remains. When Gorovitz claims that organ brokerage firms would exploit the poor, he might be assuming that no monetary value can be set upon human organs, i.e., that body parts are monetarily invaluable.[6] In this case, then, organ donors would always be exploited when they sold body organs, for with each sale they would take money in exchange for a commodity that was of inestimable financial value. The difficulty with this argument, though, is that the major premise appears false—at least when that premise is properly interpreted. It is probably true that no monetary value can be set upon human body parts, if by "monetary value" one means something like "absolute financial worth." On the other hand, monetary values can be placed upon different person's body parts if we take "monetary value" to mean "fair exchange value," and then determine that value by referring to the concrete set of circumstances in which any organ sale occurs. Let us say, for instance, that my brother needs a kidney in order to survive. Virtually everyone would agree, I think, that I would not be "exploited" or "taken advantage of" if I were to donate my kidney to save my brother's life; in this case I would receive fair value in exchange for my kidney. Now let us change the example slightly. Assume that my brother will die unless he has a heart operation costing $100,000, and that my family has no money or health insurance. If, in these circumstances, I sell my kidney for $100,000 and use that money to pay for my brother's operation, it is difficult to see how one could argue that $100,000 was not a fair price for my kidney. After all, money has value only insofar as it translates into other forms of value, and given the circumstances specified in the above heart operation example, $100,000 translates into the ability to save my brother's life. Furthermore, we have seen that my brother's life is "fair exchange value" for my kidney. Thus, if I were to sell my kidney for $100,000 in order to pay for my brother's heart operation, I would receive fair exchange value for my organ, and I would not be taken advantage of or exploited. If this is so, however, Gorovitz cannot argue that organ donors will *always* be taken advantage of when they sell their organs, for body parts are not monetarily invaluable. More specifically, fair exchange values can be placed on human organs, and although these values may vary with the circumstances in which different organ sales take place, there is no reason at all to believe that organ donors will be exploited whenever they sell their body parts. Indeed, it might well be that in the vast majority of instances, donors will receive fair exchange value for their organs.

A second reason Gorovitz opposes commercial markets in human organs is that he believes it is unjust and unsympathetic to allocate vital medical resources according to the ability to pay. For Gorovitz, need, rather than ability to pay, ought to be the primary determinate of access to medical care.[1] Now if this argument is taken seriously, it is nothing less than a call for a total reorganization of the health-care delivery system as it presently operates in the United States. A full discussion of this issue would require nothing less than a book-length commentary, and obviously, I cannot pursue a discussion of this length in the present context. This being the case, I am simply going to assume that Gorovitz is correct in supposing that need rather than the ability to pay ought to be the primary determinate of access to medical care, and then try to determine whether, given that assumption, it would be wrong to permit the buying and selling of human organs.

At present we have a number of persons who need transplantable body organs who are unable to obtain those organs. Although Gorovitz is not entirely clear on the matter, his solution to the problem seems to be something like the following:

1. Prohibit commercial markets in human organs. (This must be done because markets of this sort perpetuate the injustice of allocating vital medical resources according to the ability to pay.)
2. Allocate freely donated organs on the basis of need.
3. Attempt to change public opinion concerning organ donation, and hope that a change in national sentiment will enable us to meet our need for transplant organs wholly by use of freely donated organs.

In considering the merits of this proposal, two different timeframes must be taken into account. First, there is a stage of indeterminate length during which the attempt to change public opinion concerning organ donation and transplantation occurs. During this time it is virtually certain that freely donated organs will not meet the national need, and that some persons will suffer and die. Second, there is a time after which the movement to alter public opinion has occurred. At this time we either will have succeeded or failed in our attempt to change public attitudes; that is to say, freely donated organs will or will not be of sufficient numbers to meet our needs.

If our analysis of Gorovitz's solution to the problem of too few transplant organs is correct, it should be clear that the proposal is flawed. If people are going to suffer and die during the time we are attempting to modify public attitudes concerning donation, why should we not permit organ brokerage firms to operate during this period? Gorovitz opposes this course of action because he says it would be unjust to allow the free market to operate in this way. That is to say, organ brokerage firms would allocate vital medical resources according to the ability to pay, and so serve only the rich. But this is not true. If the rich could obtain organs from brokerage firms, it seems clear that this would release some freely donated organs for use by the poor. In short, *everyone* would benefit to some extent. Furthermore, even if it were true that only the rich would benefit from the establishment of organ brokerage firms, it is not at all clear that it would be proper to prohibit such firms from doing business. Let us say, for example, that it would take five years to change public opinion concerning organ donation, and that during this time 1000 people would die because of a lack of transplant organs. Of the 1000 people who would die, 900 are poor and middle income persons, whereas 100 are rich enough to purchase organs through brokerage firms. Let us further suppose that if organ brokerage firms were operating during the five years we were attempting to modify public attitudes concerning donation, 50 rich people would be able to purchase transplant organs and (contrary to all probability) that this would release no additional freely donated organs for use by poor and middle income persons. Given such a hypothetical situation, our choice seems simple: Permit organ brokerage firms to operate and allow 950 people to die, or prohibit these firms from operating and permit 1000 people to die. Gorovitz would have us choose the latter course of action because he feels the former is unfair. But in the situation we are now considering, we cannot help the 900 poor and middle income persons who need transplants; these people will die regardless or whether or not we allow organ brokerage firms to do business. On the other hand, if we permit brokerage firms to operate, we can save 50 people, and not to save these people simply because they are rich seems clearly immoral.

There is a second reason Gorovitz does not want organ brokerage firms to operate during the time we attempt to change public opinion concerning organ donation. Gorovitz says that "such a prohibition . . . is necessary to put to the test our capacity as a nation to meet the present shortage, and to fashion ways to deal with future shortages. . . . "[1] This just seems to be wrong. There is no reason we cannot allow organ brokerage firms to do business while we engage in a concerted attempt to change national sentiment concerning organ donation and transplantation. Indeed, it is quite likely that allowing such firms to operate would help bring about the change in public sentiment that Gorovitz so fervently desires. Gorovitz observes that there is something "distasteful" about the thought of human body parts being bought and sold[1] and in this observation he no doubt is correct. If organ brokerage firms were allowed to operate in the United States, then, this might shock the public to such a degree that a massive shift of opinion concerning organ donation would occur. And if this were to happen, free donations of organs might increase to such a degree that organ brokerage firms would lose all *raison d'etre*.

Gorovitz's final two arguments in opposition to organ brokerage firms remain to be considered. If American firms were allowed to buy and sell human organs, Gorovitz believes these firms would seek to obtain body parts from the poor in Third World countries, and that this would be wrong because it would: (i) exploit the poor in other lands, and (ii) damage our international credibility.

We have seen that there is no reason to believe that organ brokerage firms would exploit the destitute in our society; hence, we have no reason to believe that these firms would exploit the poor in other lands. This being the case, we justifiably may dismiss argument (i). On the other hand, argument (ii) contains a kernel of truth. Whether or not organ brokerage firms would, in fact, exploit the poor in other countries, it is quite likely that their actions would be perceived as exploitive of citizens in those countries. This being the case, if American brokerage firms were allowed to do business in foreign lands, our enemies would be given a powerful propaganda tool, and our international credibility could suffer irreparable damage. Should we then not permit American organ brokerage firms to do business in other lands? The question is not a simple one to answer, for it involves a comparison of quite different sorts of benefits and harms. If we permit American brokerage firms to obtain organs in foreign countries, we save additional American lives, but most likely cause damage to our international reputation. On the other hand, if we do not allow these firms to do business abroad, we save our reputation at the expense of American lives. Frankly, I do not know how to quantify or compare these consequences, and it may well be best not to allow American brokerage firms to do business in other countries. What must be stressed, however, is that even if this conclusion is accepted, it carries no weight in support of Gorovitz's claim that the operation of organ brokerage firms within the United States ought to be prohibited.

The Argument Against Prohibiting the Buying and Selling of Human Organs

If our argument thus far is correct, we must conclude that Gorovitz has not developed a compelling justification for prohibiting organ brokerage firms from doing business in the United States. In our country, the burden of proof is on those who would restrict individual liberty; even Gorovitz admits this. Thus, unless additional arguments are forthcoming, we must hold that the government would act improperly if it were to pass legislation prohibiting the establishment of a commercial market in human organs. Still, more remains to be said on the matter, for the case against proscriptive legislation is stronger than thus far indicated. That is to say, it is not simply the case that those who favor proscriptive legislation fail to carry their burden of proof. Rather, there are good reasons such legislation ought not to be enacted, and it is to some of these reasons that I would now like briefly to turn my attention.

Gorovitz has admirably staked out the main lines of attack to be used by those who oppose the establishment of a commercial market in human organs. Those who oppose organ brokerage firms cannot support their view simply by arguing that selling one's organs is potentially harmful, for our society allows its citizens to engage in numerous activities that are injurious, e.g., smoking, drinking, and so on. To show that the buying and selling of human organs ought not to be allowed, then, one must argue that there is something wrong with the buyer—seller relationship, i.e., it is coercive or exploitive, and as we have seen, it is precisely these sorts of arguments to which Gorovitz appeals. The difficulty with arguments of this kind, however, is that they rest upon assumptions that are questionable, and quite denigrating to those who will be asked to sell their organs, namely, the poor. If, for example, we argue that the poor will be coerced by financial inducements into selling their organs, we assume that people who live in grinding poverty lack autonomy, at least in regard to matters affecting their financial well-being. To deny that the poor are truly autonomous, however, not only questions their qualifications for full membership in the moral community, but also questions their ability to participate responsibly in the democratic system. Similarly, if we argue that organ brokers and/or the rich will exploit the poor when they buy their body parts, we assume either: (a) that the poor are unable fully to understand the consequences of selling their organs, or (b) that they are incapable of determining a fair value for their own body parts. Both assumptions lay the groundwork for governmental paternalism, but they do so without evidence, and at the expense of once again denigrating the poor. It is no doubt true that experimental subjects in a hospital setting sometimes have difficulty understanding informed consent forms.[7,8] In these cases, however, the subjects are in a totally unfamiliar setting, often in pain, introduced to numerous technical terms with which they have no familiarity, and so on. The case is quite different when a person is asked to sell a kidney, cornea, or some other body part. How many people fail to understand that having an operation involves pain, or that giving up a body organ involves risk? If we assume that the poor are incapable of comprehending such things, do we not once again question their qualifications for membership in the moral community?

Nor does it seem likely that the poor are unable to place a fair value on their body parts. After all, these people are in a better position than anyone else to judge the worth of their lives of what they will be given in exchange for one of their organs. Let us say, for instance, that person P's life-long dream has been to own a small business, that P has negotiated $100,000 for his or her kidney to establish such a business, that P is happy with this arrangement, and that P assures us that his or her life will be meaningful only if he or she is able to become an entrepreneur. On what possible grounds could the government claim that P's assessment of the worth of his or her kidney was wrong? Surely the government would be guilty of the rankest sort of paternalism if it were to prohibit this sale.

In sum, the case against those arguing for legislation prohibiting the buying and selling of human body parts is not simply that they have failed to construct a convincing justification for such legislation. Rather, it is that attempts to justify proscriptive legislation invoke a dangerous governmental paternalism, and do so by denigrating the poor. If the poor are to be counted as full members of the moral community, and if we are to remain true to the ideals of our democracy, we must view them as autonomous, rational beings, capable of directing the course of their own lives. And if this is so, it is they, and not the government, who should decide whether or not their organs will be sold.

Acknowledgment

I am indebted to Robert Arrington for comments on an earlier draft of this paper.

Notes and References

[1]Samuel Gorovitz (11/9/83) "Testimony: House Subcommittee on Investigation and Oversight," in this volume, pp. 3–12.

[2]For Gorovitz, the ideal solution to the present shortage of transplant organs would be to bring about "a massive shift in national sentiment about transplantation—a shift that would greatly increase participation in donation plans, and would also greatly diminish the barriers, psychological and economic, to participation by the medical profession in efficient collection and distribution of organs.."

[3]Of course, it is always possible that a dishonest broker could break the law, turn off the videotape recorder, and secretly tell a prospective kidney donor that the risk of donation was much lower than the government's "official" estimate. This being so, one might be tempted to argue that kidney brokerage firms ought to be legislatively prohibited because it is only in this way that potential donors can be *completely* protected against fraud and misrepresentation. However, there are two reasons this argument must be rejected. First, proscriptive legislation does not provide a complete protection against fraud, for if kidney brokerage firms were outlawed, underground transaction might still occur. And second, Gorovitz claims—quite rightly, I think— that the government has no business prohibiting actions merely because they could conceivably lead to abuses. Thus, the fact that some brokers *might* be dishonest does not justify prohibiting kidney brokerage firms from doing business. Rather, the proper course of action would seem to be to allow such firms to exist, and then close those that were found to be run dishonestly.

[4]Gorovitz does not argue that the poor and oppressed in our society are incapable of understanding the risks involved in organ donation when they are correctly informed regarding those risks; however, others may want to advance such an argument. I deal briefly with this issue later in this chapter.

[5]Robert Arrington (1983) Advertising and Behavior Control in *Business Ethics* (M. Snoeyenbos, R. Almeder, and J. Humber, eds.) Prometheus Books, Buffalo, NY.

[6]If this is what Gorovitz means, his argument as a whole is contradictory. That is to say, one cannot claim that the poor will be coerced into selling their organs by excessive financial inducements, and then argue that human organs are invaluable.

[7]Franz Ingelfinger (1979) Informed (But Uneducated) Consent in *Biomedical Ethics and the Law* (J. Humber and R. Almeder, eds.) 2nd Ed., Plenum, New York.

[8]John Fletcher (1979) Realities of Patient to Consent to Medical Research in *Biomedical Ethics and the Law* (J. Humber and R. Almeder, eds.) 2nd Ed., Plenum, New York.

Global Objections to Kidney Sales

A Response to Professor Humber

Samuel Gorovitz

A legal prohibition against a commercial market in transplantable organs became law on October 19, 1984, when President Reagan signed S2048, incorporating that provision of HR4080 in support of which I argued with the testimony reprinted here. Professor Humber has tried to refute the position taken in that testimony, but I find his objections unconvincing. I appreciate his providing me this opportunity to explain why.

Professor Humber claims that I go "to some pains to develop" the justification for prohibiting a commercial market in human organs. But surely he realizes that there is a substantial difference between the full development of a philosophical position and the sort of statement that is called for in the context of a public policy debate. In the 10 minutes one is allowed for the presentation of invited Congressional testimony, one would be ill advised to provide the first 10 minutes of a philosophical treatise! Instead, what is called for is the sketching of a line of argument, in language accessible to policy-makers, in favor of the position one supports. It is, thus, no proper criticism of a piece of testimony that its reasoning is enthymemic, as mine certainly is.

Professor Humber does seem to acknowledge this point late in his remarks, when he says that I have "staked out the main lines of attack to be used by those who oppose the establishment of a commercial market in human organs." That is precisely the objective of my testimony. The question then is not whether that line of argument is adequately developed in the testimony, but whether it is a defensible line of argument.

Humber correctly argues that various points I made, considered independently, fail to be conclusive. That is true of the misgivings I expressed about quality control, the possibility of adequately informing donors of the relevant risks, and the problem of potential coercion. He seems not to understand, however, that an accumulation of points, each less than decisive, can create a preponderance of evidence that has considerable weight—unless the points can each be easily dismissed. And these three points cannot be dismissed quite as easily as Humber suggests. Let us consider each of them briefly in turn.

Humber erroneously attributes to me the view that a kidney brokerage firm would face insurmountable quality control problems. My view, instead, is that quality control issues would be serious, and must not be overlooked in consideration of the plan. It is naive to think that market forces will provide adequate protection; that view is alien to the entire fabric of modern health care, in which the establishment of adequate quality control mechanisms is a major concern of patients, providers, and third-party players alike. Almost no one seriously thinks that medical transactions operate according to a classical market model.

Humber says that if many donors died or were seriously harmed, brokers would find it hard to locate new donors; thus, would the profit motive ensure high quality. But a primary concern is with the quality of the organ itself. The profit motive introduces an incentive for the prospective donor to be deceptive about having a problematic medical history, and it might then be difficult to determine the risk that the proposed organ transfer would pose to donor and to recipient. The problem is compounded when the potential donors are from populations that are typically underserved with medical care, and that do not speak or understand English. The public would have no systematic way of knowing what success rates and what incidences of harm were involved in such transactions, unless they were regulated to require long-term monitoring of donors and recipients. That would impose a degree of surveillance not typically associated with clinical interventions, but more familiar from contexts of large-scale research projects. Yet Humber claims that, "By no stretch of the imagination could [Jacobs'] firm's donors be considered research subjects!" Such regulation and monitoring could be done, of course, The point is simply that it would take considerable doing, and it would be irresponsible to dismiss such issues as trivial.

The problem of informed consent has a similar status. I have not argued that donors could not possibly be adequately informed. Ensuring that they have been so informed, however, especially given the population at issue, is considerably harder than Humber acknowledges. Barriers caused by language differences and low levels of education would make it difficult to be sure about the inadequacy of understanding on the part of potential donors. At issue, after all, is not what information is provided—that is fairly easy to regulate—but what information is actually understood.

It is puzzling that Humber balks at my reference to the regulations governing research with human subjects when he turns his attention to the problem of potential coercion. It is as if he considers respect for the autonomy of persons to be required only when they are research subjects. The requirement of informed consent is also an acknowledged constraint on clinical practice, however, and treatment without consent, except in very special circumstances, is a tort. The concerns that limit the use of research subjects apply straightforwardly to the case of potential donors; undue inducement is as inappropriate here as in the research setting.

In discussing these matters, Humber seems quite confused. He purports to offer an analysis of the concept of *coercion,* and then immediately provides an analysis of the concept of *control*. Presumably the two concepts are so obviously identical that the point need not even be noted. Yet I do not consider them to be the same at all. Control is a much more powerful notion than coercion; those who are concerned with protection against coercion are after something rather more subtle than control. Were Humber's analysis of control correct, it might help him show what I presume he means to show here—that organ brokers would not actually control prospective donors. That point I would not dispute, however, for my concern was merely with coercion. But even as an analysis of the concept of control, Humber's effort fails.

To understand Humber's account, we must accept as primitive notions the concepts of intending and ensuring. These are problematic, but I am willing to pretend that they are adequately clear for present purposes. Consider, then, the following episode. I want you to agree to come to dinner with me, so that I can bring you to the restaurant at which a surprise party awaits you. I am resolved to get you there by any means necessary, including force. But I begin by politely inviting you to join me for dinner. Thoroughly pleased, you readily accept, and accompany me to the restaurant. All the conditions of Humber's analysis of control are satisfied here, yet I have not controlled you or your behavior.

Humber's unsuccessful analysis of control appears in the context of his attempt to show that there is no real problem of potential coercion of prospective donors—sellers, actually—of kidneys. He concludes that, "There is no reason to assume that brokers will coerce customers into selling their body parts . . . or that organ brokerage firms will exploit the poor by getting them to sell their organs for less than full value." That conclusion misses the mark, however. It is not that individual brokers will be coercive that is grounds for concern. The risk is that a societal situation allowing organ sales may be coercive. This requires elaboration.

Humber provides a useful example when he asks us to contemplate his selling a kidney for $100,000 in order to finance surgery necessary to save his brother's life, and not otherwise affordable. I agree that in such a situation the sale of the kidney would be a justified action. But what a tragic situation! One can also imagine a variant of that situation. A heart patient without medical insurance could be told by a hospital that they will provide him with the needed surgery, but only on condition that he agree to let them take a kidney at the same time, for sale to another patient who needs a transplant. This variant and the original story invite the same response: No one should have to sell an irreplacable part of himself to obtain medical care—for himself or for anyone else. It is easy to describe, as Humber has done, a situation in which such transactions are justifiable. My position is simply that what is not justifiable is the toleration of such situations.

Humber claims that the story he tells shows that a "fair exchange value" of $100,000 would be established for the kidney. But the concept of a fair exchange value for a kidney makes sense only if we grant the prior assumption that kidneys are a commodity in which a market is justifiable. By Humber's reasoning, we could also establish fair exchange values for the services of prostitutes, implantable embryos, babies, or the services of contract murderers. These are all things that, under the right circumstances, some people are willing to pay for. But it does not follow from that willingness that it is justifiable that there be such a market. Humber thus begs the central question when he argues that in certain circumstances it would be rational to pay $100,000 for a kidney, and then infers that a fair market value has been established for such circumstances.

As I said in my testimony, the case for a market in kidneys is based on the reality of human desperation. It is easy to argue that, for individual pairwise exchanges, the sale and purchase of a kidney could be mutually beneficial—easy to argue, because almost surely true. The issue that one confronts at the level of public policy debate, however, is not that issue. It is, instead, the question of whether the most prudent social policy is one that allows for such exchanges because of the possibility of their local appeal. What is locally beneficial, after all, can be collectively harmful, and there can thus be good reasons to prohibit exchanges that would benefit selected individuals if they were allowed.

This should be a familiar point. For example, some people would benefit economically by being allowed to sell lead-based fuel additives to others, who would benefit by being allowed to purchase and use them. The exchange is prohibited, however, because the cumulative effects on the public health of allowing such exchanges would be detrimental. Individual liberty is thus constrained by our concern for the public welfare. But there is a limit to such constraint. So, for example, we do not adopt a repressive policy toward the leading cause of illness by imposing heavy penalties on smokers, despite the fact that the public health would benefit if we were to do so. This tension, between respect for individual liberty and concern for the public health, is a pervasive feature of health policy, and there are no conclusive arguments that will support striking one particular balance rather than another. But it should be conclusively clear that arguments about health policy cannot be resolved merely by the demonstration that maximum liberty would have its benefits for some.

The issue then becomes, once again, is it the best public policy to allow for a market in organs? Or should the needs of the poor and sick be addressed in other ways? Humber is right in attributing to me the view that medical need, rather than ability to pay, ought to be the primary determinant of access to medical care. But he is wrong in claiming that such a position "is nothing less than a call for the total reorganization of the health care delivery system as it presently operates in the United States." In fact, medical need *is* the primary determinant of access to health care in the USA today! Most people have medical coverage—through employer provided insurance, medicare, or medicaid plans, the Veterans' Administration, or one or more of several other arrangements—that will provide for their medical care primarily on the basis of need. The shame of our society is that a significant minority does not enjoy this benefit. Covering that gap would not require total reorganization, or anything like it. It would require only supplemental programs that extend protection to those presently excluded from the coverage that most of us have.

But the argument against a market for kidneys does not depend in any basic way on the fact that the poor lack equal access to health care. That problem would be eliminated at a stroke if the Federal End Stage Renal Dialysis program were extended to cover the costs of organ purchase, at market rates, for any patient in need. Distribution would then presumably be based on need. But the plan is still undesirable, not because it would fail to help anyone, not because the risks to donors or recipients could never be made clear, not because an uncoerced sale is inconceivable, but because a public health policy that allows for situations in which individuals are motivated to sell an irreplaceable part of themselves to achieve their objectives is an unwise social policy. It is unwise largely because it fails to nurture a noncompetitive, compassionate, collective response to the plight of those in desperate circumstances.

Humber claims that my proposed solution to the shortage of kidneys is to "hope that a change in national sentiment will enable us to meet our need for transplant organs wholly by use of freely donated organs." But my position is not that we should merely hope for such an outcome. It is that we ought to strive vigorously to bring it about. At this point, Humber's argument makes a crucial assumption I do not accept. He holds that we ought to allow a market at least until the donated supply is sufficient, at which point the market will fade away. That assumes that the development of a national sentiment of collective responsibility to support donation programs would be independent of the presence or absence of a commercial market. But a commercial market, even if it were otherwise unproblematic, could impede that development.

Nor is Humber's argument persuasive that while efforts are underway to increase donations, there will be needless deaths. That might be so if we were discussing heart transplants. (Perhaps Humber would argue for a commercial market in hearts, possibly limiting sales to the condemned who want to leave an inheritance and to those who have healthy hearts, but face imminent death from other causes.) But that is another topic. Those who must wait for a transplant because of the shortage of kidneys do not typically die for lack of an organ, they just continue to wait.

I have advocated that the best societal response to their circumstances is a collective resolution to provide them relief through a widespread commitment to donation of appropriate cadaver organs. I have no proof that such a happy outcome will be achieved. I have argued only that we are best advised to strive to bring it about, unimpeded by the distasteful complication of a profit-seeking commercial market in organs. I have also acknowledged that what the outcome of such an effort will be is an empirical matter, and that we would have a much better evidential base for considering a commercial market in a few years, after a serious attempt at an altruistic response. Thus, I have not argued categorically against a commercial market on absolutist grounds. I have said that the best social policy for the present is to prohibit it, and to work to bring about a state of affairs in which there is no case to be made for it. I understand that the efforts I advocate might fail. We might find ourselves, like the people in Humber's example, faced with situations in which it makes sense to allow commercial exchanges. And if those situations were widespread, and there were no better prospect of resolving them, we might have to accept the proposition that a commercial market is justifiable. Before reaching such an unhappy state of affairs, however, I prefer to work toward a society in which such situations remain hypothetical.

Other aspects of Humber's remarks are troublesome. For example, he dismisses out of hand the possibility that organs brokers would exploit the poor in other lands, despite the dismal record of some drug companies and other corporate entities in doing just that. But such points are of minor consequence, compared with Humber's fundamental misunderstanding of my position. That misunderstanding is reflected best in his claim that, "To show that buying and selling of human organs ought not to be allowed . . . one must show that there is something wrong with the buyer–seller relationship. . . . It is precisely these sorts of arguments to which Gorovitz appeals."

It is probably Humber's tendency to focus on local exchanges that leads him to such an interpretation of what is at issue. In fact, I want to argue instead on global grounds. There may well be nothing wrong with the buyer–seller relationship in particular instances of an organ sale. But there is something wrong with a society in which such an exchange can make sense. It is not my objective to prevent organ sales on the grounds that they are intrinsically evil, or that they could not ever be beneficial. What I want to prevent is a social order in which the circumstances of human desperation are best resolved by such sales. Since the cumulative effect of individually beneficial exchanges can be socially detrimental, I am unmoved by Humber's claims that organ sales could in particular cases be a good thing. I believe that prohibiting such sales, and working together toward a more caring and mutually supportive society, can lead to something better. Those are the sort of grounds on which public policy debates ought to proceed and on which a commercial market in organs has now been prohibited under Federal law.

Section II

Sex Preselection

Introduction

The section of Sex Preselection begins with an essay by Dr. Helen B. Holmes. Holmes' first step is to survey mid-1984 state-of-the-art sex preselection technology. Next, she examines the sex preferences that parents in various countries have for their offspring, and finds that preference for male children is virtually universal. Furthermore, Holmes contends that parents would probably use sex preselection procedures if such procedures were available to them. Thus, the conclusion seems obvious: If people had ready access to sex preselection technology, there would most likely be an increase in the proportion of males to females. Given this as a possibility, Holmes next discusses the effects of sex ratio imbalances, and concludes that they would probably be detrimental to females. She then considers the following three arguments in support of the rapid development and use of sex preselection technology: (1) The technology of sex preselection would be useful in improving family planning, (2) such technology would reduce suffering from sex-linked disease, and (3) sex preselection procedures would aid in controlling the population explosion. After examining these arguments in detail, Holmes concludes that none is persuasive. Finally, Holmes argues against sex preselection and claims that: (a) "If people increase masculinity and glorify it and the values associated with it, they exacerbate the traits that lead to world instability," and (b) "if individuals design particular characteristics into their children, they practice eugenics . . . [and]no human is wise enough to choose the kinds of people who ought to perpetuate our species."

Both deontological and teleological arguments have been used to support the view that sex preselection is morally wrong. The principal deontological argument is that sex preselection is an inherently sexist act, and therefore, morally objectionable. Teleological arguments, on the other hand, claim that if technologies for preselecting sex were to become widespread, numerous harmful consequences would follow. In "The Ethics of Sex Preselection," Professor Mary Anne Warren examines both sorts of argument and concludes that neither succeeds in its purpose. In response to the deontological argument, Warren claims that rational, nonsexist considerations may well prompt parents to preselect their children's sex, and if this is so, sex preselection cannot be said to be an inherently sexist act. As against the teleological arguments, Warren argues that we cannot know in advance that the net effects of sex preselection will be detrimental, and without such knowledge we cannot assert that the procedure is immoral. In the end, then, Warren concludes that until sex preselection has been practiced with some frequency and its effects are known, our presumption must be in favor of moral and legal toleration.

Sex Preselection

Eugenics for Everyone?

Helen Bequaert Holmes

Introduction

Genetic birth defects are relatively infrequent, but *every* baby has a genetic sex. If one sex were unwanted, then it could be argued that every fetus is at 50% risk for a "defect." Therefore, should sex predetermination technologies become cheap and widely used, each and every family might be making a eugenic decision for each and every pregnancy.

In this paper I shall survey briefly the mid-1980s state-of-the-art of sex determination and sex detection technologies. Thereafter, I shall (a) comment on the literature on sex preferences, (b) look at several speculations about the effects of sex ratio imbalances, and (c) consider in some detail three strong, morally based arguments for rapid development and use of such technologies. The paper ends with a discussion of the progression of John Fletcher's ethical reasoning on this topic and a composite argument *against* the selection of the sex of children.[1]

Sex Selection: The State of the Art, Mid-1980s

In humans, sex is determined at the moment of conception, when the sperm merges with the egg, usually as the egg passes down one of the fallopian tubes. Each human egg contains 23 chromosomes, one of these being the X-chromosome. Each human sperm also contains 23 chromosomes, but one of these is either an X *or* a Y. At fertilization the chromosome count is brought to 46, and either a female (XX) or a male (XY) progeny is started.

Inventing methods to interfere with or manipulate this step seems a logical maneuver. But such manipulations have been surprisingly unsuccessful. Of the three types of approaches proposed to select or favor X- or Y-sperm before fertilization, two involve technology and the third prescribes specific behaviors during coitus.

Chemical or Physical Barriers

The first suggestion is to create a barrier (chemical, such as a selective spermicide; or physical, such as a diaphragm or filter) that would allow only one type of sperm to pass the cervix for the subsequent journey through the uterus and up the fallopian tube. The publicized wish for a "manchild pill" is a plea to invent a systemic method for chemical destruction of X-sperm.[8,9] However, there has been little or no progress toward the development of either chemical or physical barriers, probably because of the very slight difference in properties between X- and Y-sperm.

In Vitro Sperm Separation

Researchers have put considerable effort—with some slight success—into a second approach: the separation of sperm in semen samples (for subsequent artificial insemination). Because one sex of a domestic animal has more commercial value than the other, research veterinarians have done most of the exploratory work in sperm separation. Two international conferences recently considered the accomplishments: In 1970 the American Society of Animal Science sponsored "Sex Ratio at Birth—Prospects for Control"; in 1982 the Warwick Land Company of Rhode Island funded "Prospects for Sexing Mammalian Sperm." Proceedings of each of these conferences have been published.[10,11]

The results reported in these two books are discouraging. One difficulty is that there is no simple way to test the accuracy of a sperm separation technology. For one test, separated sperm are killed and stained with the fluorescent dye quinacrine: the human Y-chromosome usually, but not always, contains a fluorescent "F-body." Veterinarians at such centers as the Lawrence Livermore Laboratory in California are perfecting a fluorescence-activated cell sorter, which can count and separate 1000 cells per second, for this purpose.[12]

Checking the sex ratio of progeny after artificial insemination with separated sperm is the logical and ultimately definitive method, but it is expensive. To date, progeny counts from separated bull sperm have been disappointing.[13] The claims for human sperm are more positive but are reported only by those with a vested interest in their own techniques (as one example, *see* ref. *14*.).

Ericsson and Glass recently summarized the literature on the speculated differences between X- and Y-sperm.[15] Purported differences in size, in shape, or in migration patterns to negative or positive electrodes have not been verified. Experiments reporting reactions to antisera have not been repeated successfully. As yet, there is no good experimental evidence for differential survival of the two kinds of sperm in fluids of low or high pH, although pH of a vaginal douche is a key factor in the Rorvik and Shettles method of sex preselection.[16,17] However, one fully confirmed difference exists: the X-*chromosome* is considerably larger than the Y, and therefore the total DNA from the chromosomes in an X-sperm weighs about 2.7% more than the DNA in a Y-sperm. From this fact, researchers hypothesize that X-sperm may be heavier and that Y-sperm may swim faster.

The Y-sperm's alleged "differential progressive mobility" is the basis of the "Ericsson technique," which ranch owner Ericsson first attempted as a method to separate bull sperm, and then applied to human sperm. To use this technique, clinicians place semen samples into an albumin column; the Y-sperm allegedly swim faster through the viscous liquid, and are collected to use for artificial insemination.[14,15,18–20] Ericsson holds US Patents 4,007,087, 4,009,260, and 4,339,434 on this process. According to the latest brochure from his firm Gametrics Limited, ten clinics in various USA cities and four in southeast Asia use "our semen technology to isolate sperm for sex selection (male) and/or male infertility."

But does it work? Of the first 91 children delivered at these clinics, 68 (75%) were male.[14] Except for staff members at some of the clinics, most workers in fertility are skeptical.[21] Each client does sign a detailed informed consent form that states clearly that the technique merely *increases the probability* for conceiving a boy; clients are not accepted if interviewers believe that an unwanted girl baby will be abused or aborted.

The sex selection clinic in Philadelphia is pioneering another technique, one that allegedly enriches for X-sperm.[24] Invented in 1975 by Steeno et al.,[25] improved later by Quinlivan et al.,[26] this method requires the pipeting of semen into a glass column filled with tiny beads of a gel (Sephadex gel). Fractions 5, 6, and 7 collected from the column have X-sperm enriched to 62—84% (checked by quinacrine staining). Apparently more Y-sperm than X-sperm adhere to the beads of gel and do not pass through. This procedure is so new, and so few patients have requested girl babies, that no meaningful data on results are available yet.

Coital Behavior

Many formulae for coital behavior to conceive a boy or a girl have been handed down in folklore.[27] Although one or another aspect of some of these formulae may indeed be true, no biological basis has been unequivocally demonstrated. Yet some of these methods appear in the medical literature, and twentieth century gynecologists sometimes suggest them to their patients. In America, authors apparently find it worthwhile financially to continue to write magazine articles and books on the subject.[16,17,32,33]

Bits of evidence and feasible biological hypotheses have kept some do-it-yourself methods alive in clinical circles, although skeptics greatly outnumber proponents. One theory has led to the "preconception gender diet." Several French and Canadian physicians claim to have evidence that minerals in the mother's diet can influence which sperm fertilizes her egg.[32,34—36] Biological hypotheses invented to explain clinical results obtained by the proponent physicians suggest that a woman's internal mineral balance may affect the cervical mucus through which sperm must travel, the internal surface of the fallopian tube up which sperm must swim, or the "zona pellucida" around her egg's membrane.[37]

A second theory is that a high sperm count in the female tract increases the chance that a Y-sperm fertilizes an egg. High sperm counts are found in healthy, well-nourished males who wear loose clothes around the testicles, whose mothers did not take DES during pregnancy, and who have abstained from intercourse for 2—3 days before producing the sperm sample. Several ejaculations within the same day into the same vagina apparently build up sperm count.[38]

The third theory has the most support from papers in recognized medical journals: the theory that timing of intercourse in relation to ovulation can favor X- or Y-sperm in the race to the egg.[16,17,33,40—46] However, even here experimental results from different sources produce different hypotheses; Shettles' popular books recommend timing procedures contrary to those in Whelan's book and in some of the medical articles; furthermore, some authors reverse their timing schemes for artificial insemination. If peristalsis and secretions in the female vagina affect Y-sperm mobility, then different results from artificial insemination without sexual arousal would be plausible.[47]

In their papers in the Bennett anthology,[3] James and Williamson summed up well the current status of home methods for sex selection.[45,48] James, who has probably written more papers than anyone else on the connection between timing of intercourse and sex of progeny, asserted:

> Preconceptual control of sex of infants is a topic that has attracted the attention of hoaxers, incompetents, madmen, and cranks, as well as scientists. . . . I shall call them all *sex hypothesizers*. . . . In sex hypotheses [there is] a great potential for confusion [because] the time interval between conception and delivery is so long that false predictions made at conception may be forgotten or revised at parturition (p. 73).[45]

And Williamson stated, "Sex selection techniques have been widely publicized before being tested and even those of known ineffectiveness have been touted" (p. 129).[48]

The fact that home methods of sex determination sell books and occupy space in popular magazines and medical journals without any real supporting evidence is instructive to bioethicists. The desire to determine the sex of one's child is widespread and every child is "at risk" for this trait. In advance of the discovery of cheap and accurate methods that are likely to be popular, bioethicists ought to prepare by serious consideration of the ethical issues.

Sex Detection: The State of the Art

Medicine's prying into nature's secrets about the sex of the unborn child has been much more successful. Where Western medicine is practiced, most women have come to accept many manipulations to their bodies as part of standard medical "management" during pregnancy. From prenatal medicine have come a variety of sex detection techniques, which I classify as: (a) speculative methods, (b) marginal methods (those with equivocal results or accurate only in the third trimester when the fetus is viable), or (c) essentially 100% accurate methods.

Speculative Methods

In the first scenario, a clinician removes a few cells from an in-vitro-fertilized (IVF) embryo (a "test-tube" baby), checks the sex (by staining for X- and Y-chromosomes, or by using a recombinant DNA "probe"), then implants the remaining cells only when they are of the wanted sex. (In mammals, a whole animal can usually be formed even if a few of the early divided cells are taken away.) Indeed, in rabbits, Gardner and Edwards used cells from a later stage (the blastocyst) to predict sex correctly; subsequently they succeeded in implanting and bringing to term 20% of these sexed blastocysts.[49] And in humans, a pregnancy was established (and carried for five and one-half months) from a frozen embryo with only five of its eight cells intact after freezing and thawing.[50] Deliberate removal of a healthy cell from an early human embryo for sex detection has—to my knowledge—not yet been reported.

A second scenario, again for use with IVF babies, is to determine whether the embryo will react to an antibody against a product of the Y-chromosome, either the HY- or the SDM-antigen. This method would select for females: A male embryo would clump with the antibody and could not be implanted; a female one would not clump and thus could be implanted. Although Epstein et al. claimed successful use of the technique with mice at the eight-cell stage,[51] Chapman reported that the Epstein method was unreliable.[52]

Sex selection by one or the other of these methods is sometimes proposed as a spin-off "benefit" of IVF research. If and when the rabbit and mouse results improve, and when IVF spreads to more countries, teaching hospitals, and private clinics, certainly some researchers and clinicians will surreptitiously (or openly) attempt "test-tube" sex detection.

Marginal Methods

Sex Hormone Level

Attempts have been made to detect androgens (male steroids such as testosterone) in women carrying male fetuses. Fetal steroid hormones cross the placenta and add to the hormones in the mother's blood and in her saliva. However, the mother also produces androgens; in both fetus and mother hormone production is cyclic. To date, methods of measuring androgen levels in blood in the first trimester,[53] and in saliva later in pregnancy, [54,55] have given results essentially as predictive as guessing. The popular press in Austria, Switzerland, and Germany reported prematurely the "spit test" (Speicheltest) research in those countries. In still other attempts to detect hormones produced by male fetuses, investigators have analyzed amniotic fluid, obtaining the fluid by an amniotic tap (amniocentesis). In the fluid from 60 amniocenteses, Méan et al. of Switzerland found a wide range of testosterone levels, with a statistically significant difference between the *averages* from fluid surrounding male and female fetuses.[56] However, they found so much overlap in hormone level that in 30% of the cases, no sex prediction could be made. Nevertheless, work continues with these three fluids as clinicians attempt to find new methods to predict sex reliably.

Fetal Cells in Maternal Blood

Cells from the fetus can cross the placenta; thus, clinicians have checked maternal blood samples for F-bodies (from a male fetus) with fluorescence-activated cell sorters. Reports on such attempts come from Finland, Belgium, and the USA.[57,58] The original excitement about this method seems to have faded, apparently because results differed when experiments were repeated.

Ultrasonic Visualization

Competent ultrasonographers can detect the penis or vulva in the third trimester of pregnancy except when poor images are obtained because of interference by fetal bones, breech presentation, scanty amniotic fluid, or maternal fat or bowel gas.[59,60] In one report from Australia, the genitalia were "seen" in 66% of 137 fetuses scanned at 24—40 weeks, with only a 2% error.[61] In Sweden, diagnoses were made in 74% of 101 fetuses at 32 weeks, with a 3% error.[62] Over the past five years, various medical journals have presented data from three continents that show 50—86% successful recognition of sex during the third trimester. In pregnancy management in many parts of the United States and Europe, where ultrasonography is routine in the third trimester to assess gestational age and fetal position, mothers are usually told the sex of their fetus when the body parts are clear. According to Hobbins,[63] "knowing the sex of a fetus in the third trimester is kf dubious clinical value, but may be of psychological benefit to some patients."

However, knowing the sex during the *second* trimester would be of value in decision-making about therapeutic abortion. Therefore, clinicians around the world have also attempted to use ultrasonic visualization of genitalia during that part of pregnancy. Success rates have been low; for example, Plattner et al. predicted sex in 61% of 194 fetuses at 16–24 weeks gestational age, with a 14% error,[60] and Birnholz determined sex in 41% of 367 fetuses of that age.[59] But the technology of ultrasonic scanners and the techniques of using them are constantly being improved. In a rather startling report, Stephens and Sherman of the University of California at San Francisco describe "100% accuracy of fetal anatomic-sex determination by linear-array real-time ultrasound in 100 consecutive cases of fetuses whose gestational ages ranged from 16 to 18 weeks."[64] Stephens scanned in two planes of orientation; sometimes it took as long as 10 minutes to assign sex. Should other clinicians succeed in obtaining similar results, ultrasonography may come to fall in the category below.

Essentially 100% Accurate Methods

Fetal Cells Obtained Through the Cervix

During the second month of gestation, some cells from the embryo's portion of the placenta slough off and can be found in the lower, endocervical part of the uterus. These can be obtained with a syringe in a relatively noninvasive way through the mother's cervix. Clinicians from Anshan, China, in 1975 stained such cells for F-bodies. They reported 93% accuracy in 100 pregnancies.[65]

In a recent refinement of this procedure, a tiny piece from the chorion of the embryonic placenta is aspirated through a catheter inserted via the cervix under ultrasound guidance. Recently, recombinant DNA biotechnologies have led to elegant applications of this "chorionic biopsy" technique (also called chorionic villus sampling, CVS).[66,67] A manufactured radioactive "probe" for Y-chromosome DNA can be mixed with the DNA extracted from chorionic cells. If the fetus is male, the probe will bind in two to three days in a complementary fashion to Y-chromosome DNA, a test much more accurate than the F-body stain.[68] Using this method on 13 cases in Scotland, Gosden et al. reported 100% correct sex prediction.[69] Pregnancy loss—possibly 3–4%—caused by the CVS technique itself is currently under investigation in international controlled studies.[70] However, physicians are very excited about the early diagnosis potential of this technique for use with families at risk for sex-linked disorders.

Amniocentesis

From clinical use starting in 1969, this test has now become commonplace in Western obstetrics. Invented originally to detect chromosome abnormalities, amniocentesis followed by cell culture easily identifies sex chromosomes. During the second trimester of pregnancy, a small sample of amniotic fluid is removed via a hollow needle guided through the mother's abdomen into the amniotic sac. The few fetal cells present in the fluid are coaxed into growth in tissue culture medium. When enough cells are present, sometimes after four weeks, chromosomes from dividing cells are stained and identified. Golbus et al. reported a 99.93% accuracy rate in 1979.[71] Now, in medical centers with sophisticated molecular medicine, the new recombinant-DNA probes have revolutionized this procedure also: With the DNA from a few fetal cells, and no lengthy culture procedure, the probes can identify sex in a few days. A pregnant woman and/or her physician can make an abortion decision before the woman has begun to feel the fetus move.

Preferences For Sex of Offspring

Preferences in the United States

Since 1930, the social science literature has reported more than 30 attitudinal studies or fertility behavior analyses that purport to reveal Americans' preferences for sex of offspring.[72] College students are a favorite captive experimental group for sociologists and psychologists; somewhat fewer studies have been done with pregnant women, married women, or married couples.

Despite different research designs and approaches, the results of the attitudinal studies are quite similar. As summarized by Williamson (p. 131) essentially all studies show a slight but persistent preference for boy children, combined with a wish for balance. "Americans rarely want only (or mostly) daughters . . . . The most popular combinations are: just one boy and one girl, at least one of each sex, and more boys than girls (including a single boy if only one child).[48] Pregnant women have been less willing to express sex preference than other women,[76–78] couples who already had children tended to rationalize their existing sex composition,[76] although men were less likely than women to rationalize having daughters.[79] After their recent study in Texas, Pharis and Manosevitz concluded that, although the status of women may have improved during the past decade, there is little evidence of a reduction in the preference for male babies.[77]

A strong preference that the *firstborn* be male also continues. Norman believed that he had detected a change in the preferences of students when he compared data from his research in 1974 with that done by Dinitz et al. in 1954.[81] The earlier data, obtained from about equal numbers of male and female university students, showed that 60.3% wished their first child to be male, and 34.5% had no preference.[81] Norman's data showed 48.3% wanting a firstborn male, with 45.1% showing no preference.[80] However, in 1983, among the students tested then by Pharis and Manosevitz, 62% wanted a first-born son; 32% responded "either OK."[77] Clearly over this recorded span of 30 years, Americans have maintained their explicit, albeit moderate, preference that their firstborns be male.[82]

Powledge (p. 194) strongly criticized all studies on preference for sex of offspring as "worthless because (a) they tell us that people prefer boys, which we already knew; (b) they cannot answer the question of whether the sex ratio will change, or how much: and (c) . . . they cannot help us assess the likely consequences of sex choice."[83] However, in contrast to Powledge, McClellan (p. 43) has stated, "in coming to grips with the magnitudes of the potential effects of sex-selection techniques on fertility," data yielded by surveys based on the questionnaire method are better than none.[75]

Some demographers have argued that analyses of actual reproductive behaviors, in terms of the total children in families of certain sex configurations, reveal real sex preferences. Couples whose first several children were of the same sex or predominantly female tended to have another child sooner or to have larger completed families than did other couples.[73,85–87] Such data, therefore, seem to confirm the son preference/family balance conclusions drawn from simple preference studies. However, McClelland has pointed out several problems in drawing conclusions from the configuration of sexes in families of certain sizes.[84] Subjective probabilities affect whether the family accepts the "risk" of another pregnancy. For example, a family with two girls may believe that they are now more likely to have a boy, a belief known as the *gambler's fallacy;* or another such family may believe that they are "girl producers," a belief that may have some biological basis, but is more likely to be the *trend fallacy.*[75]

Acceptance of Sex Selection Technologies in the United States

A few researchers have used "behavioral intention" measures by asking whether, if sex selection techniques were available, the interviewee would use them. In 1968, of 283 students at three Florida colleges, 26% said that they would like to choose the sex of their future children.[74] In their 1970 national fertility study, Westoff and Rindfuss found that 38.8% of 5805 currently married women responded positively to the question, "How would you feel about being able to choose the sex of a child?"[88] In 1977, Hartley and Pietraczyk analyzed 2138 responses from a random sample of students (53% of them male) in classes at five different northern California colleges.[89] They found more acceptance of the idea than the earlier workers: A majority of their respondents (65.9%) "agreed" that the technology of sex predetermination should be available to all parents; and 44.6% would want to use such procedures themselves. One could conclude that during the

1970s, a period in which medical technologies were burgeoning and copiously reported in the popular press, people became more ready to accept such technologies, whereas preferences about sex of offspring did not change.

Preferences in "Non-Western" Countries

Son preference in the United States seems trivial compared to that in most of the "developing" nations, as anthropological data clearly demonstrate. China, Korea, Taiwan, Hong Kong, Singapore, and India provide the most extreme examples; only certain ethnic groups in Thailand and the Phillipines seem to prefer family balance.[31,48] Reports on strong son preference come also from Africa.[90] Until this century, male and female infanticide was widely practiced for population control and other reasons.[39,91] However, "systematic infanticide, wherever . . . practiced [was and] is directed primarily toward females.[92] Now that explicit infanticide is prohibited by the laws of most nations, neglect or abuse causes deaths of female infants and young children.[92–94]

An illustration of this situation can be seen in India and China. In the state of Punjab in North India, a girl is a burden to a family because of the few employment opportunities for women and the large dowry that must be given when she marries. In a study of the condition of women, Horowitz and Kishwar conducted interviews in a rural village in Punjab.[95] Women often said that personally they would like to have daughters for emotional support and for help with chores. But they dread having daughters: Their life is more miserable when they produce daughters, whereas their status and treatment improve when sons are born; they suffer so much as women that they do not want to subject their own children to such misery. And the family as an economic unit will have trouble providing the dowry.[93,95]

Despite the rhetoric of equal rights for women, including many excellent laws in Indian legislation, modernization has actually exacerbated the problems of Indian women. When more male peasants were made landowners through land reform measures, the value that women once had had as equal agricultural laborers almost vanished. With mechanized threshers and grinders run by men, the hard hand labor that women used to do was no longer marketable. Son-preference became more intense and spread into other parts of India. After infanticide was outlawed in 1890 and punished by fines to the village as well as to the family, son-preference was expressed by the slow death of little girls through inadequate feeding or failure to provide medical treatment, especially for infant diarrhea. Sex ratios of 970 women for every 1000 men in 1901 fell to 930 per 1000 in 1981.[93,94,96]

Now, into this scenario came amniocentesis! Indian physicians learned the technology in the West and brought it to India. Because Indian women may legally terminate a pregnancy, it is not possible to prohibit abortions done after the results of amniocentesis are reported. Thus, this technology has come to provide a more modern and unpunishable "solution" to the daughter problem than infanticide or neglect. It has come into widespread use in India as word is spread by enthusiasts and by not-so-subtle advertising on billboards, such as "Boy or Girl? Know the sex of your unborn child . . . with the aid of . . . sophisticated scientific techniques.[94,97] The editors of the Indian feminist magazine *Manushi* decided NOT to publish a report on the extent of usage of amniocentesis for sex selection because they felt that such an exposé would boomerang by attracting more customers for the procedure![98]

For data from China, newspaper accounts are essentially the only source. The one-child policy, which started in 1979, has made parents desperate that the one child be male. The official census of 1981 recorded among some two million births that year 921 girls to 1000 boys. For children born in 1983, unofficial figures from Chinese demographers give 901 girls to 1000 boys. Apparently most little girls are eliminated by "accidents" during delivery.[99] Both chorionic biopsy and amniocentesis are available in areas with advanced medical facilities. Officially used to detect birth defects, each of these procedures is done "blind" because of the general lack of ultrasound facilities.[70]

Effects of Sex Ratio Imbalances

Those who have speculated about the social consequences of sex choice have expected that boys would be preferentially selected. In 1968, Etzioni's predictions in *Science* included: an increase in crime, "some of the rougher features of a frontier town," reduced support for the arts, and the demise of religion and moral education.[100] Since in those pre-Reagan days more women than men voted Republican, Etzioni also predicted the end of the two-party system because the proportion of Republican voters would decrease. Later Postgate, who strongly advocated sex choice to control population, also expected unpleasant consequences (but to be tolerated for the greater good of society). For example, he stated:

> It is probable that a form of *purdah* would become necessary. Women's right to work, even to travel alone freely, would probably be forgotten transiently. Polyandry might well become accepted in some societies; some might treat their women as queen ants, others as rewards for the most outstanding (or most determined) males (p. 16).[9]

For data on what actually happens when sex ratios are imbalanced, Guttentag and Secord present in the book *Too Many Women? The Sex Ratio Question* their studies of several actual modern and historical populations with imbalances.[39] They observed that societies with a preponderance of males may treat women as possessions to be bought and sold; such societies usually place emphasis on female virginity, proscribe adultery, and have a low divorce rate. Women may be forced to marry, sometimes as child brides. Also:

> Women [often are] regarded as inferior to men . . . [in] reasoned judgment, scholarship and political affairs. [They may be] excluded from any but the most elementary education (p. 79).[39]

Paradoxically, in India where women are the minority, the gender is so devalued that the scarcity of females continues to be augmented by female infanticide, girl child neglect, bride murders, and suttee (the self-cremation of Hindu widows on the funeral pyre of their husbands).

What sort of scenario did Guttentag and Secord find in populations with a preponderance of women?

> [Men] can negotiate exchanges that are most favorable to them . . . . Men are more reluctant to make a commitment to any one woman, and if they make it, it is a weaker one, and is more apt to be broken. . . . Women are apt to feel exploited, because even when they meet a male partner's demands, he may break off the relationship . . . This feeling of being exploited generates attempts by women to redefine male and female roles in a relationship, to reject a male partner, and/or to reduce their dependency by becoming more independent. (p. 190)[39]

To explain their observation that women get short shrift whether or not they are the majority sex, Guttentag and Secord formulated a social exchange theory, to which they gave major emphasis in their book. They described two forms of power in society: dyadic power and structural power. Dyadic power belongs to the sex in short supply; it determines who can "call the shots" in making dyads. However, the other form of power, structural power, is always in the hands of men and is usually the determining power. Currently in most countries women outnumber men, and men have both dyadic and structural powers. They can exploit women and always find a woman when they want one. On the other hand, in places where women are scarce, women theoretically should be determining relationships. In a very few such societies, such as the early medieval period in Europe, women did exercise some of this sort of power (p. 57). However, it is more usual for men to take tight control over women when they are a scarce resource. Daughters and wives may be kept in purdah, and/or exchanged and sold as a market commodity. When there is a hierarchy of wealth or prestige, the wealthy man may show his power by collecting more than one wife (p. 49).[39]

However, the correlation of certain social behaviors found in a variety of times and places with specific sex ratio imbalances does not necessarily indicate that the skewed sex ratios have caused those behaviors. There could be no cause-and-effect relationship. Or, particular social customs may themselves lead to, or exacerbate, unusual sex ratios, rather than the other way around. Data showing that imbalanced sex ratios tend to perpetuate themselves provide evidence that prevailing customs maintain existing sex ratios.[39]

Some of those who have speculated about bad effects of skewed sex ratios claimed that imbalances would be trivial and only temporary. Westoff and Rindfuss believed that any slight variations in sex ratio would correct themselves naturally:

> If effective sex control technologies were rapidly and widely adopted in the United States, the current sex preferences of married women indicate the the *temporary effect* would be a surplus of male births in the first couple of years. This would be followed by a *wave of female births* to achieve balance, and the oscillations would eventually damp out (p. 636). (Italics mine.)[88]

Similarly, Keyfitz stated: "The shortage of girls in the population would begin to be felt within much less than 50 years, and this would act back on the preferences of parents."[101] Such thinking assumes that parents would be motivated to bring about 50—50 sex ratios in society. The evidence from Laila Williamson's study of infanticide[91] and the data of Guttentag and Secord on the perpetuation of skewed sex ratios[39] refute such hypotheses. To change a low proportion of females in any population, the underlying low regard for females and the economic advantage of males must be changed.

However, the current preference for first-born sons means that an even lower regard and lower economic status for women might result should Americans be able to select children's sex easily and cheaply. Currently husbands are, on the average, 2.5 years older than their wives (p. 175).[39] With sex selection this gap might widen, augmenting the current power imbalance between men and women.

The many studies on firstborn characteristics concur that firstborns tend to become more distinguished and have more managerial positions than later-borns. Firstborns are likely to be "more ambitious, creative, achievement-oriented, self-controlled, serious and adult-oriented . . . more likely to attend college and to achieve eminence" (p. 93).[102] Altus found that firstborn men and women were overrepresented in elite undergraduate colleges in the United States.[103] (*see also,* Breland, Forer, and Williamson.)[73,104,105] As Pogrebin put it, "At present, at least some firstborn girls have a crack at these special advantages. But, with sex control, boys will monopolize the eldest-child bonuses in addition to other male privileges."[29]

Moral Arguments for Choosing Children's Sex

Several authors have taken strong positions in the medical, philosophical and popular literature that urge the development of effective sex selection technologies. Three views that deserve serious consideration are the arguments that sex choice will improve family planning, that it will reduce suffering from sex-linked disease, and that it will control the population explosion. Let us look at each of these.[106]

A Bonanza for Families?

The first argument might go this way: "Every child a wanted child" is a well-known goal of family planning. Some parents may believe that they cannot properly rear a child of one particular sex. A child of wanted sex may be less likely to suffer abuse or emotional or nutritional neglect. Grandparents will be more pleased. If parents get what they consider to be the ideal family constellation—that is, sexes of children in a particular order—they will be happier than with some other constellation. All children in a family will benefit from this parental satisfaction. In Bangladesh, for example, girls born into families with more boys than girls have a higher survival rate than those born into families with more girls than boys.[48] Furthermore, if parents do succeed in getting a wanted son or daughter, they may stop childbearing sooner, and the smaller family provides better economic advantages to all their children.

During the *Hard Choices* TV program "Boy or Girl? Should the Choice Be Ours?" Bill Allen, prospective father who used the Ericsson technique, said, "You know, it's simply a more sophisticated form of family planning. And why shouldn't we have the right, if the opportunity is available to us, to do that sort of planning?''[114]

The fact that parental expectations of whatever sort are often frustrated is the first rebuttal to these arguments. If the ''wanted'' boy or girl is a disappointment and does not behave as imagined, and if trouble and expense went into his/her predetermination, then he or she is at greater risk of abuse than the child of randomly determined sex.[120] And how do we decide the happy family constellation when parents disagree? If one child creates problems in the resulting family, the parent whose plan was not followed can keep the wound open. The problems in raising a family are unpredictable; the satisfactions and joys, likewise; neither of these can be spelled out in advance nor determined by sex choice.

Because sperm separation and chorionic biopsy (to be followed by selective abortion) are the two methods of sex selection most likely to be improved for use in Western nations, such sophisticated techniques would probably be available only to ''a small elite of higher-income, urban, and well-informed couples'' (p. 140)[48] Then these methods might become yet further examples of medical technologies unfairly distributed. Furthermore, if those who select firstborn males are already members of dominant groups, then those groups' advantages would become more firmly entrenched. Therefore, sex selection for family planning, like so many technological advancements, has the potential for increasing injustice.

Selection of children's sex starts us on a slippery slope. What traits would we like to be able to specify in our children? Hair color? IQ? The physique for our favorite sport? For many parents such traits in their children are more important than sex; children can be neglected or abused for failing to meet a variety of expectations. If we are going to custom-design our children, for which traits is there moral justification? There are no such traits. Any specification means that we are not genuinely interested in adding a unique person to our home. Positive eugenics—that is, the deliberate selection of genetic traits for human beings coming into this world—is morally wrong. After all, we humans are not really wise enough to know what traits will be best for the good of humankind through all eternity. Besides, we would harm the individual person we design: she or he would lose considerable freedom with the physical and mental characteristics prescribed by the humans with the power of conception. Moreover, the Nazi eugenics program clearly demonstrated the potential diabolical nature of eugenics policies. I believe with Powledge that we are blind if we think that eugenic ideas can be imposed only by governments:

> We simply do not see attempts by individual couples to achieve particular kinds of children are no different, except that the one is imposed by state power and the other appears voluntary . . . . It is not the hypothetical actions of governments that should fill [us] with trepidation, but those of the people themselves (p. 211).[116]

Preventing Sex-Linked Disease

But suppose that we wish to alleviate the suffering of children from fatal genetic diseases that can be detected by amniocentesis. Is it justifiable to eliminate all children with such diseases? This sort of eugenics (negative eugenics) may under some circumstances be morally acceptable.[121] Abortion of a fetus that has been shown with certainty to have a serious genetic or developmental defect is often in the best interests of its parents and society. The suffering of a child, such as one with Nieman-Picks disease, who wastes away in pain over five to ten years, may be eliminated. Indeed, the use of scarce resources for the palliative care of such fatal disorders as Tay-Sachs disease and microcephaly may be unjustifiable.

Evaluation of the argument for using sex selection technologies in families at risk for sex-linked diseases requires an understanding of what is meant by sex-linkage. The more correct terminology is "X-linkage." Such a disease is caused by a defective protein that is coded for by a gene located on the X-chromosome.

Females are much less likely to suffer sex-linked diseases than males for the following reason: Genes for serious defects are rare in human populations. Let us consider a hypothetical case of a certain deleterious gene that is found on 0.2% of all human X-chromosomes. Two of every thousand males would have the disease. And 0.2% females would carry the defective gene on one X-chromosome, and would *not* be sick, because they would carry a gene for a good protein on the other X-chromosome. The probability that a female would be sick with this hypothetical sex-linked disease would be 2/1000 times 2/1000, or 4/1,000,000; that is four out of every million women. With this particular gene frequency, a male is 500 times more likely to be afflicted with the disease.

An affected male always passes his defective gene to all his daughters, making all of them carriers, but never to his sons, for he gives only a Y-chromosome to each son. A family with such a father is not considered to be at risk because neither sons nor daughters will actually get the disease. But the healthy mother who carries the sex-linked gene will on the average pass that gene to half of her sons.

Therefore the family with the carrier mother is defined as the one at risk for the disease.

Currently there are no prenatal tests for the more common, serious sex-linked diseases, such as hemophilia, Lesch-Nyhan disease, and Duchenne muscular dystrophy. Therefore, in a family at risk, one would test a fetus for sex and abort any male because he has a 50% chance of being diseased. The female, with a 50% chance of being completely free of the gene and a 50% chance of being a carrier, would be allowed to come to term.

This form of negative eugenics, i.e., abortion of males who have a 50% chance of being affected, is considered to be medically (and morally?) justified by most prenatal diagnosticians. They often urge the rapid developments of sex detection technologies for this purpose.[26,63] Although I support the personal reproductive rights of parents to choose *not* to bear and raise a defective child, I cannot wholeheartedly support the use of sex detection and selective abortion for sex-linked disease. First, half the abortions are of nondefective males; second, half the survivors are girl carriers who will then face the same problem as their mothers.

Some proponents of sex selection who are not geneticists, have used a "benefit to future generations" argument to advocate sex selection in families at risk for sex-linked disease.[17] They believe erroneously that if every such family produced only daughters, the population would be rid of the disease. However, relatively few people have access to sophisticated genetic counseling; the defective gene can arise anew in the population by random mutation; and, most importantly, the many surviving carrier females may give birth to a son with the disease at any time. Indeed, the gene for a lethal sex-linked disease may actually increase in frequency when families produce daughters to compensate for the loss of sons.

However, ethical discussion about sex choice for sex-linked disease may be unnecessary in the future. Recombinant-DNA technologies are snowballing. Soon there may be created radioactive DNA probes for all of the common and many of the rare genetic diseases. Such a probe would bind to the defective gene in the DNA isolated from fetal cells obtained by amniocentesis or chorionic biopsy. When the *gene itself* can be detected, only a son who *actually has* the genetic disease need be aborted. Such developments in sophisticated Western medicine would convert ethical questions about the morality of choosing sex to avoid disease into those about prenatal diagnosis in general.

Population Control

Sex selection has been proposed as a means to control the population explosion, for example by Paul Ehrlich in *The Population Bomb* in 1968. John Postgate, a British microbiologist, writing in 1973, said:

> [T]he only really important problem facing humanity to-day is over-population . . . '[M]ultiplication in under-developed unenlightened communities is favoured, and these are the ones most prone to perpetuate the population explosion in ignorance . . . [M]y . . . panacea, one which would take advantage of such ignorance and short-sightedness . . . is a pill, or other readily administered treatment which, taken at coitus, would ensure (with . . . greater than 90% certainty) that the offspring would be male . . . . Countless millions of people would leap at the opportunity to breed male: no compulsion or even propaganda would be needed to encourage its use (p. 12, 14).[9]

Also advocating the manchild pill, Clare Booth Luce stated:

> The determining factor in the growth of all animal populations is . . . the birth rate of female offspring. Only women have babies. And only girl babies grow up to be women . . . In the overpopulated countries, the preference for males amounts to an obsession . . . . [A] pill . . . which . . . would assure the birth of a son would come as man-ah! from Heaven (p. C-1).[8]

Luce and Postgate believed overpopulation to be the most serious problem facing our planet and that drastic methods must be used to stop mass starvation. According to them the invention of a "manchild pill" would save humankind. Indeed, their logic is excellent. No compulsion to take this pill would be needed. People in countries such as India and China would gladly choose to breed boys. Once they had enough sons to establish family security and the mother's status, they might well stop reproducing. Moreover, according to Ehrlich, Luce, Postgate, and others, the real population reduction breakthrough would occur in the following generation. Since the number of babies produced depends on the number of available uteruses, one generation after widespread sex selection very few babies *could* be produced.

Postgate's article inspired some spirited negative letters to the editor of *New Scientist*, with arguments that fall into three categories. The first objection, a Kantian one, is that Postgate's plan is morally wrong because the end (solving the "only important" problem facing humanity) is being used to justify the means. "[T]he kind of standpoint . . . embodied in his thesis [is that] . . . the species *homo sapiens* must be kept going, whatever kind of creature he (she!) turns into.''[122] According to Masson, most of those in power over the centuries have subscribed to the creed that the end justifies the means, and that terrorism, torture, and a myriad other ''human-inflicted'' wrongs are defended by this creed. Sex selection as a means would corrupt the end, resulting in female slavery in one or another guise.[122] And another letter stated, ''Although mankind is the peak of God's creation, his perpetual preservation on this planet is not something I would sacrifice all else for.[123]

The second argument is that Postgate's plan is racist and classist (as well as sexist). Postgate used the terms ''unenlightened,'' ''ignorance,'' ''short-sightedness'' to describe underdeveloped countries. In his letter, Ibbett pointed out Postgate's failure to mention ''the main problem to be faced: that of the lack of any will from the rich minority of the world to sacrifice much of their affluence for the benefit of the poor majority.''[124] Prosperous citizens of the richer countries of the world have always blamed the poorer countries for threatening their opulence by breeding too many people.

The third argument from these letters is that a ''manchild pill'' might have results far worse than the intended ones. Some argued that entrepreneurs might breed daughters for financial gain, resulting in an excess of women and an acceleration of population growth.[125] And Johnston wrote, ''the women would have to be locked up under state control.''[123] The state would reward ''its panting male population . . . with . . . sex for those who serve well the party machine.'' According to Masson ''male frustration and aggression . . . would stand a good chance of destroying the species . . . , not so much in wars as in riots, raids, and drug-addiction on a vast scale.'[122] Overpopulation may not be as much of a threat to the world as the policies and behaviors that are associated with masculinity and maleness.

Therefore, sex selection as a means to cure overpopulation is likely to be pernicious. Proposing such a method is particularly ironic when existing evidence has already demonstrated that population growth slows with improvement in social welfare and extension of the roles of women beyond that of child-bearing. Family planning programs are generally unsuccessful when there is no improvement in providing the necessities of life. Birth rates are lowered with increases in income levels, health care, employment opportunities, education, and the status of women. ''The countries in which [birth rates have dropped sharply] . . . are those in which the broadest spectrum of the population has shared in the economic and social benefits of significant national progress.''[126] ''The more education women have, the fewer children they bear.''[127]

Fletcher's Ethical Analyses

John Fletcher has written on this topic more extensively than any other contemporary bioethicist. In 1979 in the *New England Journal of Medicine,* he argued that it was inconsistent in a society that (through Supreme Court decisions) permits abortion before 24 weeks for any reason, for physicians to refuse access to amniocentesis for sex choice.[128] The press distorted his view to be an ''advocacy'' of sex choice. Then several competent bioethics scholars criticized his moral arguments through letters to the *New England Journal* and responses published in the February 1980 issue of *The Hastings Center Report*. Childress and Steinfels pointed out that Fletcher had misinterpreted the Supreme Court's rulings: according to the Court, a woman has a negative right of noninterference, but not a positive right to assistance.[129,130] Childress further objected that Fletcher was appealing ''to a formal principle, consistency, rather than to a substantive principle, such as fairness, which counts for more in ethical argument.''[131]

In 1981, Fletcher reconsidered his position in a lengthy essay for the Bennett book, *Sex Selection of Children,* published late in 1983.[132] In this essay, Fletcher applied what he considers to be the dominant ethical stance of contemporary Americans toward values issues in reproduction, namely, ''freedom with fairness'':

> The limits of freedom begin when harm is inflicted upon all by its unlimited expression . . . . There are two risks to society and its institutions by disseminating the technology to make sex choice decisions. First, the unharmful desire to plan or balance children in a family could result in harm to the ideal of equality between males and females if there are significant increases in first-born males . . . . A second risk is a precedent for a reintroduction of some of the ideas of positive eugenics . . . that reflect the inordinate desire of one generation to instill its concepts of ethics and virtue in succeeding generations (p. 222, 247, 248).[132]

Despite acknowledging these risks to society, Fletcher concluded that policy-makers should not restrict sex choice technologies. He felt that if any harms were later demonstrated, ''the mills of a democratic society, strongly propelled by concepts of freedom and fairness, are probably sufficient to grind and resolve the problem'' (p. 248).[132]

Two years later, however, Fletcher,[133] reconsidered this conclusion and took a very strong stance against sex selection (except to avoid sex-linked disorders). He claimed that selection of a child's sex is unfair and sexist and that any reasons given by parents for preferring one sex can also hold for the other sex.[133] He further asserted that harmful consequences from the practice of sex selection would ''far outweigh the few fleeting beneficial consequences'' (*see* also, note *134*).

Conclusion

I concur with Fletcher's most recent position, and here extend the position, first to societies in which son preference is extreme, and then to societies like the United States.

In countries like India, China, and Korea, any available technological method for sex selection would be eagerly sought. (Since they would interfere too much with established social customs, nontechnological methods like diet and timing of intercourse would probably be less acceptable.) Because these overpopulated nations do have population-control programs, a cheap and effective method of son selection, such as a manchild pill or shot, could well be officially welcomed; methods involving selective abortion might be unapproved but condoned, as apparently is the case now in China and India.[93,94] As shown earlier, devaluation of women is often self-reinforcing; therefore, if amniocentesis should become more widely used, or if simpler technologies were available, the situation in India and China might become unbearable for women.

Under these circumstances, however, it is important not to lay blame on parents for selecting sons. Women make correct moral choices, using flawless utilitarian reasoning, when they maximize their own and their family's happiness and minimize the suffering of little girls. Before a decision to have a daughter has utility, societal practices must change so that women acquire value as persons. Women should be provided with education, meaningful employment, and the right to own land.

In Western nations, where preferences are less extreme—for firstborn sons, for ''balance,'' for a child to fill a certain role—methods involving late abortions would probably be only marginally acceptable. However, if a cheap preconception sex-choice method really worked, or if chorionic biopsy became easily available to permit selective early abortions, couples might well act on their preferences. Probable outcomes would include increased inequality between the sexes, a curtailment of women's freedom, more glorification of stereotypically masculine traits, and a variety of negative social effects from the higher proportion of firstborn males.

But will a simple method become available soon? In 1968, Etzioni reported scientists' estimates that routine sex control of humans would be available in seven to 15 years.[100] Now more than 15 years have passed and no such methods have been invented. There is good reason to think that simple sex control is unattainable in principle for these reasons: In animals and plants various methods of sex determination have evolved. Through natural selection, mechanisms for maintaining a highly advantageous sex ratio have also evolved for each species. Therefore, only clever and involved technological mechanisms will succeed in interfering in this sex determination process. There are complicated technical difficulties in developing a shot or pill that would kill all the X-bearing sperm in a male without otherwise harming him and/or the Y-sperm, or in developing a vaginal suppository that would kill one kind of sperm in an ejaculate. And in the laboratory, scientists still cannot cleanly separate the two kinds of sperm from each other.[22]

The most feasible foolproof method for wide adoption in Western nations (and probably eagerly snatched by other nations) is the chorionic biopsy. As yet, however, its safety and quality controls have not been worked out, although much research attention is focused here.[70] Though this procedure seems simple, it qualifies as high technology because of the sophisticated cell culture and/or recombinant-DNA biotechnologies needed to make it useful. It may be expensive and perhaps not permitted for sex selection in some places.

It might here be asked: if no methods are feasible, is this paper a case of tilting at windmills? No. It is important for bioethicists to stay aware of current progress in the various technologies and to examine the arguments that have been used to urge rapid development of sex selection. I have shown here that reasons for promoting sex selection may be unethical. Problems that sex selection allegedly can solve may instead be solved in morally acceptable ways, and, in fact, might be aggravated if sex selection, or even the mindset necessary for sex selection, were prevalent.

Finally, I wish to reemphasize two great dangers intrinsic in the pro-sex-control mindset. Each danger jeopardizes the survival of humankind. First, if people increase masculinity and glorify it and the values associated with it, they exacerbate the traits that lead to world instability. Second, if individuals design particular characteristics into their children, they practice eugenics: No human is wise enough to choose the kinds of people who ought to perpetuate our species. There may be some things that we *can* do, but that we *ought not* to do: Perhaps sex selection is one of them.

Acknowledgments

I am grateful to Robert Almeder, Diana Axelsen, Francis Holmes, and Jodi Simpson for their thoughtful suggestions and criticisms of an earlier version of this chapter, and especially to Betty B. Hoskins for the benefit of our extensive previous collaboration. However, any errors and all opinions remain my responsibility.

Notes and References

[1]Two recent collections of essays have dealt with interdisciplinary issues raised by sex preselection. For the first, which emerged from the 1979 workshop, "Ethical Issues in Human Reproduction Technology: Analysis by Women," Janice Raymond and Emily Culpepper have coordinated three papers by feminist scholars, followed by three responses and a lively audience discussion.[2] In the more recent anthology,[3] twelve authors from six disciplines have provided bibliographic information through 1981 and have discussed the issues competently from a variety of perspectives. Recommended earlier interdisciplinary reviews are those by Largey, Rinehart, and Williamson.[4–7]

[2]Janice Raymond, section ed. (1981). Sex Preselection in *The Custom-Made Child? Women-Centered Perspectives. (Helen B. Holmes, Betty B. Hoskins, and M. Gross, eds.)* pp. 177–224. Humana, Clifton, New Jersey.

[3]Neil G. Bennett, ed. (1983) *Sex Selection of Children.* Academic Press, New York.

[4]Gale Largey (1972) Sex control, sex preferences, and the future of the family. *Social Biology* 19, 379–392.

[5]Gale Largey (1973) Sex control and society: A critical assessment of sociological speculations. *Social Problems* 20(3), 310–318.

[6]Ward Rinehart (1974) Sex preselection—not yet practical. *Population Report,* series I, #2. pp. I-21–I32. The George Washington University Medical Center, Washington, DC.

[7]Nancy E. Williamson (1976) *Sons or Daughters: A Cross-Cultural Survey of Parental Preferences.* Sage, Beverly Hills.

[8]Clare Boothe Luce (1978) Next: Pills to make most babies male. *Washington Star* July 9, C-1–C-4.

[9]John Postgate (1973) Bat's chance in hell. *New Scientist* 58(841), 12–16.

[10]C. A. Kiddy and H. D. Hafs, eds. (1971) *Sex Ratio at Birth—Prospects for Control*. American Society of Animal Science, Champaign, Illinois.

[11]Rupert P. Amann and George E. Seidel, Jr., eds. (1982) *Prospects for Sexing Mammalian Sperm*. Colorado Associated University Press, Boulder, Colorado.

[12]Barton L. Gledhill, Daniel Pinkel, Duane L. Garner, and Marvin A. Van Dilla (1982) Identifying X- and Y-Chromosome-Bearing Sperm by DNA Content: Retrospective Perspectives and Prospective Opinions, in *Prospects for Sexing Mammalian Sperm*, (Rupert P. Anmann and George E. Seidel, Jr., eds.) pp. 177–191. Colorado Associated University Press, Boulder, Colorado.

[13]Robert H. Foote (1982) Functional Differences Between Sperm Bearing the X- or Y-Chromosome in *Prospects for Sexing Mammalian Sperm* (Rupert P. Anmann and George E. Seidel, Jr., eds.) pp. 212–218. Colorado Associated University Press, Boulder, Colorado.

[14]F. J. Beernink and R. J. Ericsson (1982) Male sex preselection through sperm isolation. *Fertility and Sterility* 38(4), 493–495.

[15]Ronald J. Ericsson and Robert H. Glass (1982) Functional Differences Between Sperm Bearing the X- or Y-chromosome in *Prospects for Sexing Mammalian Sperm*, (Rupert P. Anmann and George E. Seidel, Jr., eds.) pp. 201–211. Colorado Associated University Press, Boulder, Colorado.

[16]David Rorvik and Landrum B. Shettles (1970) *Your Baby's Sex: Now You Can Choose*. Dodd, Mead, Toronto.

[17]Landrum B. Shettles and David Rorvik (1984) *How to Choose the Sex of Your Baby* Doubleday, New York.

[18]A. Adimoelja, R. Hariadi, I.G.B. Amitaba, P. Adisetya, and Soeharno (1977) The separation of X- and Y-spermatozoa with regard to the possible clinical application by means of artificial insemination. *andrologia* 9(3), 289–292.

[19]W. Paul Dmowski, Liliana Gaynor, Ramaa Rao, Mary Lawrence, and Antonio Scommegna (1979) Use of albumin gradients for X and Y sperm separation and clinical experience with male sex preselection. *Fertility and Sterility* 31(1), 52–57.

[20]Ronald J. Ericsson (1977) Isolation and storage of progressively motile human sperm. *andrologia* 9(1), 111–114.

[21]See the conference discussion on page 219 of Amann and Seidel[11] and the remarks by Foote.[13,22] Furthermore, recently staff at Michael Reese Hospital in the clinic licensed by Gametrics Limited has seemed less than enthusiastic.[23] Also Ericsson and Glass themselves were skeptical about all other sperm separation techniques except their own.[15]

[22]Robert H. Foote (1982) Prospects for Sexing: Present Status, Future Prospects and Overall Conclusions in *Prospects for Sexing Mammalian Sperm*, (Rupert P. Anmann and George E. Seidel, Jr., eds.) pp. 285–288. Colorado Associated University Press, Boulder, Colorado.

[23]Ramaa, Rao, Mary Lawrence, and Antonio Scommegna (1983) Use of

albumin columns in sperm separation for sex preselection (Abstract). *Fertility and Sterility* 40(3), 409.

24 Stephen L. Corson, Frances R. Batzer, and Sheldon Schlaff (1983) Preconceptual female gender selection. *Fertility and Sterility* 40(3), 384–385.

25 O. Steeno, A. Adimoelja, and J. Steeno (1975) Separation of X- and Y-bearing human spermatozoa with the Sephadex gel-filtration method. *andrologia* 7, 95–97.

26 W. Leslie G. Quinlivan, Kathleen Preciado, Toni Lorraine Long, and Herlinda Sullivan (1982) Separation of human X and Y spermatozoa by albumin gradients and Sephadex chromatography. *Fertility and Sterility* 37(1), 104–107.

27 For entertaining accounts, see Guttmacher and Hechinger; Pogrebin; Rorvik and Shettles; Sangari; and Williamson.[16,17,28–31]

28 A. F. Guttmacher and G. Hechinger (1961) Old wives' tales about having a baby. *Parents Magazine* (Oct), 184.

29 Letty Cottin Pogrebin (1981) Bias Before Birth in *Growing Up Free: Raising Your Child in the 80's*. pp. 81–101. Bantam Books, New York.

30 Kumkum Sangari (1984) If You Would Be the Mother of a Son in *Test-Tube Women: What Future for Motherhood?* (Rita Arditti, Renate Duelli Klein, and Shelley Minden, eds.) pp. 256–265. Routledge and Kegan Paul, London.

31 Nancy E. Williamson (1978) *Boys or Girls? Parents' Preferences and Sex Control*. Population Bulletin. Population Reference Bureau, Washington, DC.

32 Sally Langendoen and William Proctor (1982) *The Preconception Gender Diet*. M. Evans, New York.

33 Elizabeth Whelan (1977) *Boy or Girl?* Bobbs-Merrill Company, Indianapolis.

34 J. Lorrain and R. Gagnon (1975) Sélection préconceptionnelle du sexe. *L'Union Médicale du Canada* 104, 800–803.

35 J. Stolkowski and J. Choukroun (1981) Preconception selection of sex in man. *Israel Journal of Medical Sciences* 17, 1061–1066.

36 J. Stolkowski and M. Duc (1977) Rapports ioniques: $(K^+/Ca^{2+} + Mg^{2+})$ et $(K^+ + Na^+ /Ca^{2+} + Mg^{2+})$ dans l'alimentation de femmes n'ayant que des enfants du même sexe. Enquête rétrospective. *L'Union Médicale du Canada* 106, 1351–1355.

37 Charles H. Debrovner (1982) Foreword in *The Preconception Gender Diet* (Sally Langendoen and William Proctor) pp 5–8. M. Evans, New York.

38 "The Talmud says that if a man wants all of his children to be sons, he should cohabit twice in succession."[39] On pages 100–111, Guttentag and Secord have cited sources and evaluated several biological theories.

39 Marcia Guttentag and Paul F. Secord (1983) *Too Many Women? The Sex Ratio Question*. Sage Publications, Beverly Hills.

40 Rodrigo Guerrero (1974) Association of the type and time of insemination within the menstrual cycle with the human sex ratio at birth. *N. Eng. J. Med.* 291(20), 1056–1059.

[41]Rodrigo Guerrero (1975) Type and time of insemination within the menstrual cycle and the human sex ratio at birth. *Studies in Family Planning* 6, 367–371.

[42]Susan Harlap (1979) Gender of infants conceived on different days of the menstrual cycle. *N. Eng. J. Med.* 300(26), 1445–1448.

[43]William H. James (1971) Cycle day of insemination, coital rate, and sex ratio. *Lancet* i, 112–114.

[44]William H. James (1976) Timing of fertilization and sex ratio of offspring—A review. *Ann. Hum. Biol.* 3(6), 549–556.

[45]William H. James (1983) Timing of Fertilization and the Sex Ratio of Offspring in *Sex Selection of Children.* (Neil Bennett, ed.) pp. 73–99. Academic Press, New York.

[46]Sophia J. Kleegman (1954) Therapeutic donor insemination. *Fertility and Sterility* 5(1), 7–31.

[47]Guttentag and Secord have listed flaws in published studies, and have attempted to explain away discrepancies in reports about the timing of intercourse, citing most of the pertinent literature.[39] Also, Shettles and Rorvik (p. 70–79)[17] and James (p. 73–80)[45] evaluated theories that confirm or compete with theirs.

[48]Nancy E. Williamson (1983) Parental Sex Preferences and Sex Selection in *Sex Selection of Children.* (Neil Bennett, ed.) pp. 129–145. Academic Press, New York.

[49]R. L. Gardner and R. G. Edwards (1968) Control of the sex ratio at full term in the rabbit by transferring sexed blastocysts. *Nature* 218, 346–348.

[50]Alan Trounson and Linda Mohr (1983) Human pregnancy following cryopreservation, thawing and transfer of an eight-cell embryo. *Nature* 305, 707–709.

[51]C. J. Epstein, S. Smith, and B. Travis (1980) Expression of H-Y antigen in preimplantation mouse embryos. *Tissue Antigens* 15, 63–67.

[52]Verne M. Chapman (1982) Gene Products of Sex Chromosomes in *Prospects for Sexing Mammalian Sperm,* (Rupert P. Amann and George E. Seidel, Jr., eds.) pp. 115–117. Colorado Associated University Press, Boulder, Colorado.

[53]Allan R. Glass and Thomas Klein (1981) Changes in maternal serum total and free androgen levels in early pregnancy: Lack of correlation with fetal sex. *Am. J. Ob. Gyn.* 144, 656–660.

[54]K. R. Held, U. Burck, and Th. Koske-Westphal (1981) Pränatale geschlectsbestimmung durch den GBN-speicheltest. Ein vergleich mit den Ergebnissen der pränatalen chromosomendiagnostik. *Geburtshilfe und Frauenheilkunde* 41, 619–621.

[55]von K. Loewit, H. G. Kraft, and W. Brabec (1982) Zur geschlechtsbestimmung des fetus aus dem speichel der mutter. *Wiener klinische Wochenschrift* 94, 223–226.

[56]M. Méan, G. Pescia, D. Vajda, J. B. Pelber, and G. Magrini (1981) Amniotic fluid testosterone in prenatal sex determination. *Journal de Génétique humaine* 29(4), 441–447.

[57]L. A. Herzenberg, D. W. Bianchi, J. Schröder, H. M. Conn, and G. M. Iverson (1979) Fetal cells in the blood of pregnant women: detection and enrichment by fluorescence-activated cell sorting. *Proc. Natl. Aca. Sci. (USA)* 76, 1453–1455.

[58]M. Kirsch-Volders, E. Lissens-Van Assche, and C. Susanne (1980) Increase in the amount of fetal lymphocytes in maternal blood during pregnancy. *J. Med. Gen.* 17, 267–272.

[59]Jason C. Birnholz (1983) Determination of fetal sex. *N. Eng. J. Med.* 309, 942–944.

[60]Gunther Plattner, Wilhelm Renner, John Went, Laura Beaudette, and Gilles Viau (1983) Fetal sex determination by ultrasound scan in the second and third trimesters. *Ob. Gyn.* 61, 454–458.

[61]Lachlan Ch. deCrespigny and Hugh P. Robinson (1981) Determination of fetal sex with ultrasound. *Med. J. Aust.* 2, 98–100.

[62]E-M. Weldner (1981) Accuracy of fetal sex determination by ultrasound. *Acta Ob. Gyn. Scand.* 60, 333–334.

[63]John C. Hobbins (1983) Determination of fetal sex in early pregnancy. *N. Eng. J. Med.* 309, 979–980.

[64]John D. Stephens and Sanford Sherman (1983) Determination of fetal sex by ultrasound. *N. Eng. J. Med.* 309, 984.

[65]Tietung Hospital (1975) Fetal sex prediction by sex chromatin of chorionic villi cells during early pregnancy. *Chin. Med. J.* 1, 117.

[66]Virginia Cowart (1983) First-trimester prenatal diagnostic method becoming available in U.S. *J. Am. Med. Assoc.* 250(10), 1249–1250.

[67]Michel Goossens, Yves Dumez, Liliana Kaplan, Mieke Lupker, Claude Chabret, Roger Henrion, and Jean Rosa (1983) Prenatal diagnosis of sickle-cell anemia in the first trimester of pregnancy. *N. Eng. J. Med.* 309(14), 831–833.

[68]J. R. Gosden, C. M. Gosden, S. Christie, H. J. Cooke, J. M. Morsman, and C. H. Rodeck (1984) The use of cloned Y chromosome-specific DNA probes for fetal sex determination in first trimester prenatal diagnosis. *Hum. Gen.* 66, 347–351.

[69]J. R. Gosden, A. R. Mitchell, C. M. Gosden, C. H. Rodeck, and J. M. Morsman (1982) Direct vision chorion biopsy and chromosome-specific DNA probes for determination of fetal sex in first-trimester prenatal diagnosis. *Lancet* ii, 1416–1419.

[70]Bernadette Modell (1985) Chorionic villus sampling: Evaluating safety and efficacy. *Lancet* i, 737–740.

[71]Mitchell S. Golbus, William D. Loughman, Charles J. Epstein, Giesela Halbasch, John D. Stephens, and Bryan D. Hall (1979) Prenatal genetic diagnosis in 3000 amniocenteses. *N. Eng. J. Med.* 300(4), 157–163.

[72]In 1976, Williamson thoroughly reviewed the sex preference studies and then summarized her review in later papers.[7,48,73] Largey, Markle and Nam, and especially Rinehart presented fairly complete bibliographies of the early preference studies.[4,6,74] McClelland has given a recent critical analysis of types of studies and what can be gained from them.[75]

[73]Nancy E. Williamson (1976) Sex preferences, sex control, and the status of women. *Signs: Journal of Women in Culture and Society* 1, 847–862.

[74]Gerald E. Markel and C. B. Nam (1971) Sex predetermination: Its impact on fertility. *Soc. Bio.* 18, 73–82.

[75]Gary H. McClelland (1983) Measuring Sex Preferences and Their Effects on Fertility in *Sex Selection of Children.* (Neil Bennett, ed.) pp. 13–45. Academic Press, New York.

[76]E. Eckard (1978) Sex Preference for Children and its Relationship to Current Family Composition, Intent to Have More Children, and Important Demographic Characteristics: Provisional Results From the National Survey of Family Growth, Cycle II. Paper presented at the Annual Meeting of the Southern Regional Demographic Group, San Antonio, Texas.

[77]Mary E. Pharis and Martin Manosevitz (1984) Sexual stereotyping of infants: Implications for social work practice. *Social Work Research and Abstracts* 20, 7–12.

[78]Roberta Steinbacher (1984) Sex Choice: Survival and Sisterhood. Paper presented at the Second International Interdisciplinary Congress on Women, Groningen, the Netherlands. Also in *Man-Made Woman* (Genoveffa Corea, Renate Duelli Klein, Jalna Hanmer, Helen B. Holmes, Betty B. Hoskins, Madhu Kishwar, Janice Raymond, Robyn Rowland, and Roberta Steinbacher, eds.) Hutchinson Education, London 1985, in press.

[79]Gerald E. Markle (1974) Sex ratio at birth: Values, variance, and some determinants. *Demography* 11, 131–142.

[80]Ralph D. Norman (1974) Sex differences in preferences for sex of children: A replication after 20 years. *J. Psych.* 88, 229–239.

[81]S. Dinitz, R. R. Dynes, and A. C. Clarke (1954) Preference scales for number and sex of children. *Population Studies* 29, 273–298.

[82]For anecdotal information on firstborn son preference in jokes and from transcripts during childbirth, see Pogrebin (p. 84, 87–88).[29]

[83]Tabitha Powledge (1981) Unnatural Selection: On Choosing Children's Sex in *The Custom-Made Child? Women-Centered Perspectives* (Helen B. Holmes, Betty B. Hoskins, and M. Gross, eds.) pp. 193–199. Humana, New Jersey.

[84]For example, some studies combined the effect of number preference with that of sex preference, and some surveys were based on spot decisions given to an interviewer, without a chance for spouses to confer. Furthermore, responses may have reflected what the interviewee thinks the interviewer ought to hear (impressions managment), as Steinbacher believed after her study with Faith Gilroy in which 59% of women pregnant for the first time expressed "no preference" for the sex of that child.[78] For more detailed descriptions of faulty experimental methods, see McClelland and Powledge.[75,83]

[85]D. S. Freedman, R. Freedman, and P. K. Whelpton (1960) Size of family and preferences for children of each sex. *Am. J. Soc.* 66, 141–146.

[86]Elmer Gray and N. Marlene Morrison (1974) Influence of sex of first two children on family size. *J. Heredity* 65, 91–92.

[87]Charles F. Westoff, R. B. Potter, Jr., and P. C. Sagi (1963) *The Third*

Child: A Study in the Prediction of Fertility. Princeton Univ. Press, Princeton.

[88]Charles F. Westoff and Ronald R. Rindfuss (1974) Sex preselection in the United States: Some implications. *Science* 184, 633–636.

[89]Shirley Foster Hartley and Linda M. Pietraczyk (1979) Preselecting the sex of offspring: Technologies, attitudes, and implications. *Soc. Bio.* 26(2), 232–246.

[90]Elmer Gray, Valina K. Hurt, and S. O. Oyewole (1983) Desired family size and sex of children in Nigeria. *J. Heredity* 74, 204–206.

[91]Laila Williamson (1978) Infanticide: An Anthropological Analysis in *Infanticide and the Value of Life* (Marvin Kohl, ed.) Prometheus, Buffalo, New York.

[92]Barbara D. Miller (1981) *The Endangered Sex: Neglect of Female Children in Rural North India*. Cornell Univ. Press, Ithaca. pp. 44–48.

[93]Madhu Kishwar (1984) Amniocentesis for Female Feticide in India in the Context of a Growing Deficit of Women. Paper presented at the Second International Interdisciplinary Congress on Women, Groningen, The Netherlands. Also in *Man-Made Woman* (Genoveffa Corea, Renate Duelli Klein, Jalna Hanmer, Helen B. Holmes, Betty B. Hoskins, Madhu Kishwar, Janice Raymond, Robyn Rowland, and Roberta Steinbacher, eds.) Hutchinson Education, London 1985, in press.

[94]Viola Roggencamp (1984) Abortion of a Special Kind: Male Sex Selection in India in *Test-Tube Women: What Future for Motherhood?* (Rita Arditti, Renate Duelli Klein, and Shelley Minden, eds.) pp. 266–277. Routledge and Kegan Paul, London.

[95]Bernard Horowitz and Madhu Kishwar (1982) Family life—the unequal deal. *Manushi* 11, 2–18.

[96]I was unable to examine directly the Indian census data. There is a slight discrepancy between the figures reported by Kishwar and by Roggencamp. However, demographers consider even the ratio found in 1901 extreme. Indeed, since World War II, in essentially all countries except India, China, and Korea, women have outnumbered men. For example, in the United States in 1975 there were 1024 women for every 1000 men (p.15).[39]

[97]Arun Chacko (1982) Too many daughters? India's drastic cure. *World Paper*, November, 8–9.

[98]Madhu Kishwar (1984) Personal communication.

[99]Michael Weisskopf (1985) China's crusade against children. *The Washington Post National Weekly Edition*, 28 Jan. 6–9.

[100]Amitai Etzioni (1968) Sex control, science, and society. *Science* 161, 1107–1112.

[101]Nathan Keyfitz (1983) Foreward in *Sex Selection of Children* (Neil Bennett, ed.) pp. xi–xiii Academic Press, New York.

[102]N. Lauerson and S. Whitney (1977) *It's Your Body: A Woman's Guide to Gynecology*. Grosset & Dunlap, New York.

[103]William D. Altus (1965) Birth order and academic primogeniture. *J. Personality and Social Psychology* 2(6), 872–76.

[104]H. M. Breland (1973) Birth order effects: A reply to Schooler. *Psychological Bull.* 10(3), 86–92.

[105]L. Forer (1977) *The Birth Order Factor.* Pocket Books, New York.

[106]Bioethical analyses on this topic are less common than those on other issues in reproductive medicine. Gale Largey provided the entry for the *Encyclopedia of Bioethics.*[107] Holly Goldman analyzed specifically amniocentesis as a method of sex selection.[108] Janice Raymond raised some issues from a feminist perspective in her contribution to the interdisciplinary collection she compiled.[109] Roberta Steinbacher, a psychologist, raised questions and took some firm feminist positions in a series of papers.[110–112] The subject was one of those considered by the Genetics Research Group of The Hastings Center, codirected by Tabitha Powledge and John Fletcher[113] and discussed briefly in five short articles in the February 1980 issue of *The Hastings Center Report.* Sex selection was chosen as one of six topics for the NOVA TV Series *Hard Choices.*[114] In his book *Reproductive Ethics,* Michael Bayles devoted 4½ pages to the subject depending heavily on the arguments presented in the Raymond collection.[2,115] The most extensive work is that of John Fletcher, Tabitha Powledge, Helen Holmes, and Betty Hoskins. Fletcher's evolution of ethical reasoning is described later in this chapter. Powledge, calling sex preselection "the original sexist sin," presented well-reasoned utilitarian and deontological arguments.[83,116] Hoskins and Holmes in several papers contributed "feminist values analyses."[117–119]

[107]Gale Largey (1978) Reproductive technologies: Sex selection. *Encyclopedia of Bioethics:* 1439–1444. Macmillan, New York.

[108]Holly S. Goldman (1980) Amniocentesis for sex selection. *Ethics, Humanism, and Medicine,* pp. 81–93. Liss, New York.

[109]Janice Raymond (1981) Sex Preselection: A Response in *The Custom-Made Child? Women-Centered Perspectives* (Helen B. Holmes, Betty B. Hoskins, and M. Gross, eds.) pp. 209–212. Humana, Clifton, New Jersey.

[110]Roberta Steinbacher (1980) Preselection of sex: The social consequences of choice. *The Sciences,* 20, 6–9, 28.

[111]Roberta Steinbacher (1981) Futuristic Implications of Sex Preselection in *The Custom-Made Child? Women-Centered Perspectives* (Helen B. Holmes, Betty B. Hoskins, and M. Gross, eds.) pp. 187–191. Humana Press, Clifton, New Jersey.

[112]Roberta Steinbacher (1984) Sex Preselection: From Here to Fraternity in *Beyond Domination: New Perspectives on Women and Philosophy* (Carol Gould, ed.) pp. 274–282, Romwan and Allenheld, Totowa, New Jersey.

[113]Tabitha Powledge and John Fletcher (1979) Guidelines for the ethical, social, and legal issues in prenatal diagnosis. *N. Eng. J. Med.* 300, 168–172.

[114]PBS (1981) Boy or girl? Should the choice be ours? *Hard Choices* (TV Series). Transcript available from PTV Publications, P.O. Box 701, Kent, OH 44240.

[115]Michael D. Bayles (1984) *Reproductive Ethics.* Prentice-Hall, New Jersey.

[116]Tabitha Powledge (1983) Toward a Moral Policy for Sex Choice in *Sex*

Selection of Children (Neil Bennett, ed.) pp. 201–212. Academic Press, New York.

[117]Helen B. Holmes (1976) Sex Preselection: A Feminist Perspective. Paper presented at "Assembly on the Future," Rochester, NY.

[118]Helen B. Holmes and Betty B. Hoskins (1984) Preconception and Prenatal Sex Choice Technologies: A Path to Femicide? Paper presented at the International Interdisciplinary Congress on Women, Groningen, The Netherlands. Also in *Man-Made Woman* (Genoveffa Corea, Renate Duelli Klein, Jalna Hanmer, Helen B. Holmes, Betty B. Hoskins, Madhu Kishwar, Janice Raymond, Robyn Rowland, and Roberta Steinbacher, eds.) Hutchinson Education, London 1985, in press.

[119]Betty B. Hoskins and Helen B. Holmes (1984) Technology and Prenatal Femicide in *Test-Tube Women: What Future for Motherhood?* (Rita Arditti, Renate Duelli Klein, and Shelley Minden, eds.) pp. 237–255. Routledge and Kegan Paul, London.

[120]An expectation of any sort lays a heavy burden on a child. Sons may be driven to succeed in materialistic ways. Once a perceptive student told me that, although she was constantly reminded through her childhood that her parents had really wanted her to be a boy, she felt that in her family the burden on a wanted boy would have been unbearable. According to Steinbacher, for a daughter, "the psychological ramifications subsequent to the discovery that one was chosen-to-be-second are unmeasurable but predictably negative.[112]

[121]One problem with negative eugenics is, of course, how to define "defect." Should we define one sex as a defect to be eliminated because it would suffer under current social conditions?

[122]David I. Masson (1973) Letter to the editor. *New Scientist* 58(841), 121.

[123]Jos. Johnston (1973) Letter to the editor. *New Scientist* 58(844), 305.

[124]Peter Ibbett (1973) Letter to the editor. *New Scientist* 58(841), 121.

[125]G. S. Cardona (1973) Letter to the editor. *New Scientist* 58(841), 121.

[126] William Rich (1973) *Smaller Families Through Economic and Social Progress*. Monograph #7. Overseas Development Council, Washington, DC.

[127]Kathleen Newland (1979) *The Sisterhood of Man*. Norton, New York.

[128]John C. Fletcher (1979) Ethics and amniocentesis for fetal sex identification. *N. Eng. J. Med.* 301, 550–553. Also in *The Hastings Center Report* 10(1), 15–17.

[129]James C. Childress (1980) Prenatal diagnosis for sex choice—Negative and positive rights. *The Hastings Center Report* 10(1), 18–19.

[130]Margaret O'Brien Steinfels (1980) Prenatal diagnosis for sex choice: The Supreme Court and sex choice. *The Hastings Center Report* 10(1), 19–20.

[131]John C. Fletcher (1983) Is sex selection ethical? in *Research Ethics*. pp. 333–348 Liss, New York.

[132]John C. Fletcher (1983) Ethics and Public Policy: Should Sex Choice be Discouraged? in *Sex Selection of Children* (Neil G. Bennett, ed.) pp. 213–252. Academic, New York.

[133]See Warren's paper in this volume for a clear exposition of and criticism of the argument presented by Fletcher[131] and Bayles[115] that sex preference is irrational because any reasons for preferring one sex can hold for the other sex.

[134]Dr. McIntyre of Case Western Reserve University reluctantly decided to perform amniocentesis for sex detection for a family with two daughters that would come into a million dollar inheritance if they had a son to carry on the family name.[114] The sexism in the will is obvious. Was Dr. McIntyre's decision right? It was probably not possible for him to examine the will. Could the money have gone to one of the daughters if she kept her maiden name and passed that name to her children? And what would be the disposition of the money otherwise? If it would go to a well-managed charitable or educational institution, perhaps that would be more just than to have it go to one family.

The Ethics of Sex Preselection

Mary Anne Warren

Introduction

In the chapter in this volume, Sex Preselection: Eugenics for Everyone?, Dr. Helen B. Holmes argues that it is morally wrong to preselect the sex of one's children, or even to wish to do so. (She does not, however, believe that it ought to be legally banned.) In the first part of the article, Holmes provides a comprehensive overview of the current state of the art of sex preselection. She then summarizes the rather depressing facts about the prevalence of son-preference throughout the world: In almost every culture, the majority of prospective parents, women as well as men, would rather have a male firstborn child (or only child, if they plan to have just one) and/or more males than females. Finally, she presents a number of moral arguments against the development and use of either pre- or postconceptive methods of sex preselection. Since Holmes has done an excellent job of presenting the factual background, I will confine my comments to the moral arguments.

Holmes' view is that sex preselection is a sexist act, although it is sometimes inappropriate to blame individual parents for their desire to preselect their children's sex. Moreover, she argues that if the use of new medical technologies for preselecting sex were to become widespread, the consequences for women would probably be harmful. I will argue that, on the contrary, sex preselection is not necessarily a sexist act, though it may be so in many instances. Furthermore, I doubt that it is possible to know in advance what the long-term social consequences of sex preselection will be, or that these consequences will be, on balance, detrimental to women or society as a whole. That there is a risk of harmful consequences is enough to justify continued research and monitoring of the social and psychological effects of sex preselection; but it does not justify a wholesale condemnation of the practice.

Is Sex Preselection Sexist?

Sexism is usually defined as wrongful discrimination on the basis of sex. Discrimination based on sex may be wrong either because it is based on false and invidious beliefs about persons of one sex or the other, or because it unjustly harms those discriminated against. For now, let us concentrate upon the claim that sex preselection is sexist because it is invariably motivated by sexist beliefs.

Tabitha Powledge presents the argument for this claim in its simplest form. In her view, sex preselection is "the original sexist sin," because it makes "the most basic judgment about the worth of a human being rest first and foremost on its sex (p. 197)."[1] In this form, her argument is unsound; it is false that all persons who would like to preselect the sex of their children believe that members of one sex are inherently more valuable. Some people, for instance, would like to have a son because they already have one or more daughters (or vice versa), and they would like to have at least one child of each sex. Others may believe that, because of their own personal background or circumstances, they would be better parents to a child of one sex than the other. On the surface, at least, such persons need not be motivated by any invidious sexist beliefs. They may well believe that women and men are equally intelligent, capable, and valuable; they may even be feminists, dedicated to the elimination of restrictive sex roles and sexist discrimination of all sorts.

It may, however, be argued that the desire to preselect sex is always based on covertly sexist beliefs. Michael Bayles notes that the desire for a child of a particular sex is often instrumental to the fulfillment of other desires, such as the desire that the family name be carried on. Such instrumental reasons for sex preference, he argues, are always ultimately based upon irrational and sexist beliefs. For instance, in many jurisdictions it is no longer true that only a man can pass his family name to his children; hence, he says, it would be irrational (in those jurisdictions) to prefer a son for this particular reason. Even the desire to have a child of each sex is, according to Bayles, irrational, because there are no valid reasons for supposing that this would be better than having several children of the same sex. He considers the case of a man who already has two daughters and would like to have a son as well, "so that he could have certain pleasures in child-rearing—such as fishing and playing ball with him (p. 35)".[2] This man would be making a sexist assumption, since he could perfectly well enjoy such activities with his daughters.

John Fletcher also argues that the desire to preselect a child's sex (except for certain medical reasons) can only be based on irrational and sexist beliefs. Holmes apparently agrees with Fletcher's conclusion:

> *Prima facie* examination of any argument for sex selection cannot overcome the unfair and sexist bias of a choice to select the sex of a child. The desire to control the sex of a child is not rational, since any claim that is made for the parents' preference for one sex can be demonstrated to be provided also by the other sex (p. 347).[3]

Fletcher is not opposed to sex preselection when it is done in order to avoid the birth of a child with a sex-linked disease, such as hemophilia. Women who are genetic carriers of a sex-linked disease often choose to abort male fetuses because males, unlike females, will have about a 50% chance of suffering from the disease. This is not a sexist reason for preselecting sex, although, as Holmes points out, even this use of sex preselection has some morally troubling aspects. (For one thing, it requires the abortion of some perfectly normal male fetuses; for another it entails the birth of some female children who are themselves carriers of the genetic disease.) The question is whether there are any other nonsexist reasons for sex preselection.

Holmes speaks of the situation of women in the rural parts of northern India. The society is a harshly patriarchal one, in which the birth of a male child is celebrated, but the birth of a female is regarded as a severe misfortune. Son-preference is traditionally so strong that, up to about the end of the last century, the members of some tribes killed virtually all of their female infants.[4] Although infanticide is no longer openly practiced, female children still have a higher death rate than males, because they are more often neglected, underfed, or denied essential medical care.[5] Women in this society sometimes say that they are reluctant to bring a female child into a world in which she will be abused and devalued, as they themselves have been. Holmes notes that their preference for sons would seem to be morally correct on utilitarian grounds. I would add that their son preference is not necessarily a sign of sexism on their part. To accuse such women of sexism because they act upon their understanding of the intense sexism of their society would be a case of blaming the victim. Their motivations are at least partly altruistic, and do not appear to be in any way irrational. Thus, although the use by such women of selective abortion or other methods of sex selection to produce sons must be seen as a symptom of sexist institutions and ideology, there is not necessarily anything in its motivation that would justify calling it a sexist action—even one for which the women in question are personally blameless.

Another highly pragmatic reason for son-preference in northern India (and many other parts of the world) is that a son is an economic asset, whereas a daughter usually is not. Because of sexist discrimination in the job market, a daughter will almost certainly earn far less than a son. If the family is well-to-do, she is apt not to enter the job market at all. Thus, she will not be able to contribute as much to her family's economic support. Furthermore, the cost of providing her with a dowry is likely to be extremely high. Without a large dowry she will probably be unable to marry, and thus, will be apt to remain dependent upon her family indefinitely. If she does marry without a dowry that is considered suitably large (or, indeed, even with such a dowry), she may be tormented or murdered by her husband or inlaws. Under these conditions, it would be difficult to show that the desire to have sons rather than daughters is irrational. It would surely be wrong to condemn the decision of a couple not to have children because they judge that they cannot afford to raise them. Why, then, should we condemn their decision not to have daughters, for the same reason?

If son-preference is rational in rural Punjab, and not necessarily a sexist action, then it will be difficult to argue that this is not also true in much of the rest of the world. Wherever son-preference is especially pronounced, it is because, in large part, of powerful economic motivations. Even in societies that provide some social support for the aged, sons are often an important part of old-age security—more so than daughters, whose earning capacity is generally far less. For this reason, son-preference is often (though not always) stronger among the poorer classes. Giurovich, for instance, found that son-preference is stronger among lower-class Italian couples, primarily because sons are seen as more conducive to the family's upward economic mobility.[6] Even among (some groups of) Americans, son-preference has been found to correlate negatively with socioeconomic class, suggesting that here, too, economic factors may be among the motivations for it (p. 131).[7]

It will not do to argue, as Fletcher does, that such economic motivations for son-preference are irrational because, "Few jobs exist that women cannot perform as well or better than men when performance is the criterion for evaluation (p. 343)."[3] Although this is certainly true, the fact remains that women's average earning capacity is far from commensurate with the true value of their work. As everyone knows, the average full-time employed woman in America earns just 59% of what the average man earns, and the average woman with a college degree still earns less than the average man with only a high school education. Poor women, especially if they have children, have few opportunities of escaping poverty. The morally appropriate social response to this situation would be to remove the economic incentives for son-preference through such measures as the elimination of unjust discrimination against women in education, hiring, and promotion, the provision of more adequate unemployment, old-age, and disability support for all persons, and the reduction of economic differentials through a more just distribution of wealth. But until such social changes occur, it is not necessarily irrational for poor people to seek to better their economic status through the preselection of sons.

Is it nevertheless immoral for them to do so? It might be argued that in opting for sons for economic reasons, parents are, in effect, seeking to exploit the sexism of their society for their own economic gain. Yet we cannot condemn their actions for this reason alone, unless we are also prepared to condemn the actions of women who earn a living through (for instance) modeling in bikinis for soft drink commercials. Such women may profit from sexist attitudes and institutions, but they are more often victims than victimizers; and they often have very few economically comparable options. If their actions, or those of parents who preselect sons for economic reasons, are immoral, it can only be because of their unintended social consequences.

Before turning to the consequentialist arguments against sex preselection, I would like to consider some other apparently nonsexist reasons for sex-preference. Even in the industrialized nations, prospective parents may have sound reasons to prefer that their children, for their own sake, be male. Women are still far from enjoying the full range of freedoms and opportunities available to men. On the average, they not only earn much less, but also work longer hours, because regardless of whether they have jobs, they are still expected in most cases to shoulder heavier domestic responsibilities. Male violence and the threat of male violence still turn the lesser size and upper-body strength of females into a serious liability. The threat of rape still curtails women's freedom of movement. As long as these many forms of sexist oppression persist, I think that it is wrong to suggest that women are performing a sexist action if they seek to have male children in order that the latter may enjoy the freedoms that women are still denied.

I am not, of course, suggesting that most women reason in this way; still less that most women ought to. Many prospective mothers would be equally content with a child of either sex; and many others would prefer to have a daughter. Of these, some are planning to raise a child without a male partner and believe that under the present conditions they would have more in common with a child of their own sex, and thus, (they hope) a better relationship with her. A son could share most of their particular interests or activities, but he could not share the basic experience of being female in a society that still values males more highly. However much he may sympathize with the plight of women, he will still be a member of the more privileged sex. Although such expectations may prove mistaken in particular cases, I see no grounds for condemning them as either sexist or irrational.

Other women may prefer to have daughters because they fear that, in Sally Gearhart's words,

> . . . if they have sons, no amount of love and care and nonsexist training will save those sons from a culture where male violence is institutionalized and revered. These women are saying, "No more sons. We will not spend twenty years of our lives raising a potential rapist, a potential batterer, a potential Big Man (p. 282)."[8]

Men, as a group, are far more apt to resort to serious violence against other persons (and, for that matter, against nonhuman animals) than are females. We need not speak of war, into which men are often conscripted against their will; it is enough to glance at the statistics on individual acts of violence. In the United States, for instance, males commit five times as many murders as females.[9] Rape and child molesting are primarily (though not exclusively) male crimes, and most battered spouses are female victims of male violence. The question is whether it is morally wrong to take account of such proven statistical differences between the sexes in deciding whether and how to make use of the new methods of sex preselection.

Most feminists would agree that it is usually unjust to discriminate against individuals of either sex on the basis of merely statistical differences between the sexes. Individuals have the right to be judged on their own merits, not condemned by association with some group to which they happen to belong. But choosing to have a daughter rather than a son, on the grounds that females tend to be less violent, is not a case of injustice against an individual person. The son one might have had instead might or might not have turned out to be violent, but since he does not exist, there is no way to evaluate him as an individual. Furthermore, since he does not exist he cannot have been treated unjustly; he will not suffer from his nonexistence. This is most clearly true when preconceptive methods of sex selection are used. But even sex-selective abortion cannot be regarded as an injustice against an individual person, because, as I have argued elsewhere, fetuses are not yet persons and do not yet have a right to continued existence.[10,11]

Consequentialist Objections

Numerous speculations have been made about the long-term consequences, should an effective means of sex preselection become widely available. Some writers have welcomed sex preselection as a voluntary means of reducing the birth rate and the number of unwanted children born in the attempt to get one of the "right" sex. Others, including Holmes and Fletcher, argue that the results are likely to be primarily detrimental. They fear that females may be psychologically harmed by the implementation of son-preference. Equally disturbing are the possible social consequences of sex ratio increases, i.e., increases in the relative number of males. An undersupply of women might result in their being increasingly confined to subordinate "female" roles and/or subjected to increased male violence. Let us look first at these possible negative results of sex preselection.

Birth Order Effects

Throughout most of the world, a majority of prospective parents would prefer a male firstborn child. And firstborns, it has often been claimed, enjoy certain social and psychological advantages, perhaps because they have, for a time, a monopoly on their parents' attention. There have been hundreds of studies purporting to prove or disprove linkages between birth order and such personal traits as initiative, creativity, anxiety, affiliation, dependence, conservatism, rebelliousness, authoritarianism, mental illness, criminality, and alcoholism. The results are enormously complex and frequently contradictory. However, among the most consistent findings are that firstborns tend to achieve more in terms of formal education and career, and to be more dependent and affiliative (p. 411).[12] Adler argued that each birth order position carries with it characteristic advantages and disadvantages. In his view, firstborns tend to be more responsible, conservative, and achievement-oriented, but may also suffer from anxiety and other mental problems because of the traumatic experience of "dethronement" by the birth of a younger sibling.[13,14,15] Robert Zajonc has argued that firstborn children tend to be more intelligent than laterborns because of the progressive degradation of the family's "intellectual environment" supposedly produced by the birth of each additional child.[16,17]

If either of these theories about the psychological effects of birth order were empirically well supported, there might be good reason to fear that increases in the relative number of male firstborns will have a detrimental effect upon women. However, the evidence for these theories is, at best, highly ambiguous. The isolation of birth order effects from the effects of socioeconomic status, ethnicity, religion, family size, urban versus rural background, and other social variables represents an extremely difficult methodological problem, and one that has not been resolved in most of the studies that have been done. In many of the early studies that found firstborns to be superior in intelligence, motivation, and achievement, there were no controls for family size. Obviously, firstborns are more apt to come from small families than laterborns. In most of the industrialized nations, parents of large families tend to have less money and education and to score lower on standard tests of intelligence than parents of smaller families. Thus, where family size is not held constant, comparisons between first- and laterborn children are biased. The latter have, on the average, less privileged socioeconomic backgrounds, and any psychological differences found are as likely to be a result of this factor as birth order itself. Where sample groups of first- and laterborns are matched for family size and socioeconomic status, most (though not quite all) of the apparent superiorities of the firstborn disappear (p. 45).[18]

The birth order debate continues, with some psychologists presenting new evidence of the influence of birth order upon personality and others debunking the idea. But at present, the weight of the evidence seems to support a sceptical view. In 1983 two Swiss psychologists, Cecile Ernst and Jules Angst, published an exhaustive review of the birth order research of the past four decades. Their conclusion is that nearly all of the reports of significant birth order effects are a result of errors in the design of studies and the statistical analysis of the data (p. 13).[18] They believe that there are some general differences in the socialization process undergone by firstborns and laterborns, i.e., firstborns tend to be better cared for in infancy and to be more advanced in linguistic development. But they conclude that these differences "do not seem to leave indelible traces that can be predicted (p. 187)." They do not deny that being a first, middle, last, or only child may have great importance for the personal development of some individuals, but only that it has any general and lasting significance. Birth order theories, they argue, ignore the fact that each child has a unique genetic constitution that influences his or her intelligence and personality, and that consequently, "each child in a sibship will interact in a novel way with the environment, and, from the first day on, will mold it and be molded by it in a highly individualistic way (p. 242)."

Other Psychological Harms

What about the fact that many females will know that their parents chose to have sons first? Roberta Steinbacher asks, "What are the implications of being second born, and knowing at some early age that you were planned-to-be-second (p. 187)?"[19] It might seem self-evident that girls will suffer a loss of self-esteem from the knowledge that their parents chose to have a son first. And Fletcher argues that even a firstborn girl is apt to be damaged if she learns that, whereas she was not sex-selected, her younger brother was. Addressing the parent who already has a daughter and is considering a sex-selected son, Fletcher says,

> . . . put yourself in your daughter's place. How will she respond to your reasons why you went to the fertility clinic to start a pregnancy with baby brother, when you did not do the same with the conception of her? What reasons will you give her? . . . You would not let her continue believing that only boys can be police, firefighters, or surgeons, would you? . . . You conclude that if you would not neglect her need to aspire equally to almost any job that a man might do, you will sabotage that parental duty by preselecting sex (p. 343).[3]

This argument rests upon the assumption that there can be no nonsexist reasons for preselecting sex—or none that a female child can be expected to understand. But surely one need not believe that only boys can grow up to be police or firefighters to want a son as well as a daughter. One might, for instance, believe that children are apt to develop a better understanding of persons of the opposite sex if they have an opposite-sex sibling. Or one might believe that the best way to raise a nonsexist child is to raise her in the company of an opposite-sex sibling whom one does not treat any differently. Even if these beliefs are false, they are not obvious instances of sexism. Nor is it obvious that a girl would be apt to suffer psychological harm as a result of learning that her parents preselected a son because of such beliefs.

But what of the girl who learns that her brother was sex-selected for reasons that are sexist, e.g., because her father wants a male heir? No doubt she may be hurt by this knowledge. Yet, if her parents are sexist in their current behavior, if they treat her as worth less than her brother because of her sex, then the discovery that his sex was preselected can only come as one more confirmation of what she must already know. On the other hand, if her parents are not biased in their current treatment of her brother and her, then this particular discovery need not shake her conviction that she is equally valued—although, of course, it might. Every female must eventually come to terms with the sexist biases of her society. It would be difficult to prove that the implementation of son-preference through sex preselection will do much extra damage to female psyches.

Sex Ratios and the Status of Women

Very few studies have been made of the relationship between sex ratios and sex roles. The only well-developed theory in this area is that presented by Marcia Guttentag and Paul Secord.[20] Guttentag and Secord argue that women tend to be disadvantaged in both high and low sex ratio societies—although not necessarily any more so than in societies with a 50:50 sex ratio. On their theory, high sex ratio societies tend to impose rigid restrictions upon the sexual behavior of women and to confine them to the domestic role. Low sex ratios, on the other hand, tend to contribute to male misogyny and the devaluation of both women and marriage. This is because when there is an "oversupply" of women, men become reluctant to commit themselves to long-term relationships with a single woman. In such circumstances, women are apt to become dissatisfied with the terms of marriage, and to seek other means of achieving economic security; hence, feminist movements may appear. According to Guttentag and Secord, whichever sex is in short supply is likely to gain an advantage in "dyadic power," i.e., power within two-person heterosexual relationships. Yet men are usually able to limit women's freedom even when sex ratios are low, because they have the advantage in "structural power," i.e., control of the economic, legal, and other key social institutions.

On this theory, high sex ratios may be either good or bad for women, depending upon the structural power that women already have. If women are economically dependent and lack basic legal rights and protections, they cannot make use of whatever dyadic power they might otherwise gain as a result of their scarcity value. But if they have a degree of economic independence and legal autonomy, then they may be able to take advantage of this dyadic power to drive a better bargain in relationships with men. This benefit applies primarily to heterosexual women. Guttentag and Secord say very little about nonheterosexual women, except that lesbianism is apt to be more severely discouraged in a high sex ratio society. Insofar as compulsory heterosexuality is a basic part of the oppression of women, this must be seen as an additional danger of sex ratio increases.

However, as Holmes points out, we cannot assume that the differences between typical high and low sex ratio societies are actually caused by sex ratios. It may be that the causal relationship tends to run in the opposite direction. Rather than high sex ratios causing women's confinement to the domestic role, societies that confine women to the domestic role may tend to have high sex ratios because many parents conclude that raising females is less worthwhile than raising males, and therefore practice sex selection through female infanticide or the neglect of female children. The high sex ratios may be relatively harmless in themselves. They might, conceivably, even be beneficial to women, in the context of a society that allows them very few opportunities to lead a decent life outside of the wife-and-mother role.

Even if it is true that in the past high and low sex ratios have tended to have the social consequences that Guttentag and Secord describe, it would be a mistake to assume that in the future the results will be the same. One may speculate that women in the more severely patriarchal societies will be apt to suffer a further loss of freedom should sex preselection lead to a significant increase in sex ratios; without substantial structural power, women cannot benefit from their own increased "value." Yet nothing in this scenario is inevitably predetermined. Improved education and movements toward socialism and democracy tend to facilitate the loosening of traditional constraints upon women, and might tip the balance in favor of sexual egalitarianism even in the face of declining sex ratios. The power of women's liberation movements throughout the world is another unpredictable factor. I suspect, however, that the growth of mass communication will make it increasingly difficult for women to be kept ignorant of their own oppression and the need to struggle against it.

Women in the more industrialized and/or less severely patriarchal nations probably have somewhat less need to fear increased oppression as a result of sex ratio increases. Not only is the relative increase in the number of males apt to be smaller, but because women often have greater (though still inadequate) opportunities for economic independence and political influence, they may be able to successfully defend those rights already won, while continuing to improve their legal and economic status. This is not predetermined either; the forces of reaction are strong, and the gains that women have already made may be lost, with or without sex ratio increases. Nevertheless, sweeping predictions of a loss of freedom for women should sex ratios increase are unjustified.

Are We on a Slippery Slope?

Another consequentialist objection to sex preselection is that it may lead to the predesigning of children in other respects than sex. Holmes notes that some parents may be more concerned about a child's hair color or IQ than about its sex. She asks, "if we are going to custom design our children, for which traits is there moral justification?" Her reply is that, "There are no such traits. Any specification means that we are not genuinely interested in adding a unique person to our home." There are two strands of argument here. One is a version of the slippery slope argument: We should reject sex preselection because it will lead to other forms of positive eugenics, which are objectionable. The other is that we should reject all forms of positive eugenics, because any attempt to predesign a child indicates a refusal to treat that child as a unique individual. Both arguments are questionable.

All slippery slope arguments presuppose that people cannot (learn to) make certain distinctions that the arguer considers vital; if the relevant distinctions can be made, then there is no reason to suppose that acceptance of the one form of behavior will lead to acceptance of the other. Such arguments fail if either (1) people can make such distinctions, or (2) these distinctions do not have the significance that the arguer takes them to have. In this case, both conditions apply. Many people who do not object to sex preselection would object to the preselection of hair color or IQ, because they perceive that these cases involve quite different considerations. Most of the arguments for and against sex preselection would not normally apply to the preselection of hair color, which usually has much less social significance than sex. Preselecting for intelligence would raise much more serious moral questions, because intellectual ability has a much more direct effect upon a person's life prospects than hair color normally does. These questions can and must be treated separately.

I am puzzled by the suggestion that all forms of positive eugenics are indicative of an unwillingness to perceive a child as a unique individual. Positive eugenics includes all attempts to select for certain traits that are positively desired, as opposed to selecting against certain undesirable traits, such as hemophilia or Down's Syndrome. Positive eugenics tends to evoke the image of dictatorial governments predesigning people to serve their own nefarious ends, or of parents predesigning children to fit their own entirely selfish preferences. It is easy to forget that some forms of positive eugenics might serve children's own interests. Suppose, for instance, that there was a perfectly safe preconceptive or prenatal procedure that would endow a child with excellent vision or an increased life expectancy. I can see no *a priori* reason to deny that such a procedure might provide a real benefit to future persons. Why should we assume that parents who wish to provide their children with such benefits are uninterested in adding a unique individual to their home? As in the case of sex preselection, their reasons might be selfish or irrational, but they might also be altruistic and well-reasoned.

Positive eugenics may be feared for a number of reasons: It might be abused for immoral purposes; it might prove to have unforeseen side effects; it might divert medical resources from more important purposes; and, above all, the advocacy of eugenics has been historically associated with vicious racist doctrines. These are sound reasons for proceeding cautiously, with full public disclosure and extensive public discussion of each new or proposed procedure—just as should be done in every other area of medical technology. They are not, however, reasons for a blanket rejection of all such procedures. Many possible eugenic procedures will prove too expensive or too dangerous to be worth pursuing. But each must be evaluated on its own merits. If we refuse to make the essential distinctions, insisting that all forms of positive eugenics must be accepted or rejected as part of a single package, then we may inadvertently contribute to the very sorts of abuses that we fear.

Will More Women Be Born Poor?

Some observers have predicted that if sex preselection becomes readily available, the poorer classes will become increasingly male, since son-preference is often strongest among the most economically deprived (p. 1109).[21] On the other hand, if the new methods of sex preselection continue to be expensive, or if governments, fearing their social consequences, seek to ban their use, sex preselection may become a prerogative of the relatively wealthy. In that case, it will probably be the upper classes that experience the greatest increase in sex ratios. As Steinbacher points out, this would mean "that increasing numbers of women in the future are locked into poverty while men continue to grow in numbers in positions of control and influence (p. 188)."[19]

This is perhaps the most damning of all the consequentialist objections to sex preselection. The detrimental effects of a further "masculinization of wealth" would be difficult to overestimate. Increased wealth and power in the hands of men could only result in the aggravation of the entire range of injustices against women. Yet we cannot move directly from this fact to the conclusion that the development and use of sex preselection is morally objectionable. What is morally objectionable is that it should be made available only to the wealthy. If we want to avoid some of the worst social consequences of sex selection, we must either suppress it completely (which is probably impossible), or seek to make it equally available to all social classes. It is much too early to predict that the latter goal will prove impossible.

The Possible Benefits

The possible benefits of sex preselection are just as difficult to predict as are the possible harms. I agree with Holmes that sex preselection should not be lauded as a means of reducing the birth rate. We cannot be sure that fewer children will be born if parents are able to preselect sex; some parents may have more children if they can be assured that they will be of the preferred sex. Nor do we know that decreases in the relative number of women will have the effect of reducing birth rates. In those cases in which pronatalism remains strong, high sex ratios may only result in each woman being expected to have more children. If a shortage of women were to result in polyandrous marriages, women who received support from several men might find it possible—and perhaps necessary—to have more children than would be feasible in a monogamous marriage. Fertility drugs might even be used to increase the number of multiple births.

A more realistic possibility is that governments would take steps to prevent sex selection from resulting in a severe shortage of women. Whereas any absolute prohibition would probably be unpopular and ineffective, a variety of less severe measures might be employed. Couples might be forbidden to use sex preselection to produce sons until they have already produced at least one daughter; or tax penalties or other disincentives might be used to reduce the attractiveness of all-male families. Ways might even be found to reduce economic discrimination against women, thus reducing son-preference. Thus, the long-term effect of sex selection upon birth rates is quite unpredictable.

Moreover, as Holmes points out, there are better ways of promoting voluntary reductions in the birth rate than by encouraging the use of sex selection to produce sons. Free universal access to contraception and abortion is essential (and nonexistent in much of the world), but will be insufficient unless combined with more far-reaching social reforms. Improved economic security, education, and health care, and expanded opportunities for women outside of the maternal role have consistently proven effective in lowering birth rates. These measures are desirable on independent moral grounds, and should be supported even by those who doubt that overpopulation is a real problem.

I also agree with Holmes that we cannot be certain that parents will be happier if they are able to choose the sex of their children. No doubt some will be happier and others will only be disappointed when their sex-selected children fail to live up to their expectations. Getting what one wants is never a guarantee of happiness—although it is usually more conducive to happiness than not getting what one wants.

There are, however, two predictable benefits of sex preselection that do much to counterbalance its possible ill effects. The most important is that fewer children will be doomed to abuse or neglect because they are of the "wrong" sex—in most cases, because they are female. It is true that even a wanted girl or boy may suffer from unrealistic parental expectations. But wanted children are less likely to be deliberately deprived of food, affection, and necessary medical care; and fewer wanted children die from such neglect. We will never know how many short and miserable lives will be avoided through sex preselection, but the data on differential mortality rates for female children in northern India and many other parts of the world suggest that the number will be quite significant. In my mind, this potential benefit is at least as weighty as any of the potential harms that Holmes describes. I doubt that any of the possible benefits to be gained through discouraging the development and use of new methods of sex preselection is worth condemning even a few children to rejection and neglect.

Sex preselection will also provide at least some women with a new means of resistance to patriarchy. It is part of the oppression of women that they have generally had little choice but to bear and raise sons, thereby perpetuating the ruling sex/class. Women may soon have the option of refusing to do this, without avoiding motherhood altogether or abandoning their male children, as the legendary Amazons were said to do. Other women, less optimistic about the prospects for change, may resist patriarchy by refusing to add to the female underclass. The freedom to preselect the sex of one's children is far less vital to women's interests than the freedom to decide whether to bear a child or not; yet having the former option will still be important to some women. Granted, some women may be forced by their husbands or families to have sons when they would prefer daughters, just as some are forced to complete pregnancies they would prefer to abort, or to abort those they would prefer to complete. But the option of sex choice will still have value for those women with the desire and the opportunity to use it.

Conclusion

I have not argued that the net effects of sex preselection are bound to be beneficial. They may well prove to be detrimental, just as Holmes fears. My primary point is rather that we cannot possibly know in advance what the effects of sex preselection will be, and that we ought not to condemn it on the basis of what can be little more than speculation. Were it possible to prove that sex preselection is, in every instance, a sexist act, then it could be condemned without proof of a high probability of serious harm. But if, as I have argued, there are many nonsexist reasons for son-preference or daughter-preference, then sex preselection can be morally condemned only if the consequentialist arguments against it are very strong. Because these arguments are not particularly strong, because there are probable compensatory benefits as well as possible ill effects, and because the possibility of net losses does not justify categorical condemnation, the presumption must be in favor of moral, as well as legal, toleration. Should the feared detrimental effects of preselection begin to materialize at some future time, then will be the time to reassess this moral stance.[22]

Notes and References

[1]Tabitha Powledge (1981) Unnatural selection: On choosing children's sex. in *The Custom-Made Child? Women-Centered Perspectives.* (Helen B. Holmes, Betty B. Hoskins, and Michael Gross, eds.), Humana, New Jersey.

[2]Michael B. Bayles (1984) *Reproductive Ethics.* Prentice Hall, New Jersey.

[3]John C. Fletcher (1983) Is sex selection ethical? *Research Ethics,* Alan R. Liss, New York.

[4]Kanti B. Pakrasai (1970) *Female Infanticide in India* Editions India, Calcutta.

[5]Barbara D. Miller (1981) *The Endangered Sex: Neglect of Female Children in Rural North India* Cornell University, Ithaca and London.

[6]G. Giurovich (1956) Sul desiderio dei coniungi di avere figle e di avere figle di un data sesso. (On the wish of married couples to have children and to have children of a specified sex.) *Atti Della 16 Riunoine Scientifica della Societa Italiana di Statistica,* Rome.

[7]Lee Rainwater (1965) *Family Design* Aldine, Chicago.

[8]Sally Gearhart (1982) The future—if there is one—is female. in *Reweaving the Web of Life: Feminism and Nonviolence.* (Pam McAllister, ed.), New Society, Philadelphia.

[9]National Commission on the Causes and Prevention of Violence (1969) *Violent Crime.* George Brazillen, New York.

[10]Mary Anne Warren (1973) On the moral and legal status of abortion. *The Monist* 57(1), 43–61.

[11]Mary Anne Warren (1977) Do potential people have moral rights? *Can. J. Phil.* 7(2).

[12]Bert N. Adams (1972) Birth order: A critical review. *Sociometry* 35(3), 411–439.

[13]Alfred Adler (1927) *Understanding Human Nature* Greenberg, New York.

[14]Alfred Adler (1928) Characteristics of the first, second, and third child. *Children,* 3(14).

[15]Alfred Adler (1931) *What Life Should Mean to You* Little Brown, Boston.

[16]Robert B. Zajonc (1975) Dumber by the Dozen. *Psychol. Today* (January), 37–43.

[17]Robert B. Zajonc, Hazel Markus, and Gregory B. Markus (1979) The birth order puzzle. *J. Pers. Soc. Psychol.* 37(8), 1325–1341.

[18]Cecile Ernst and Jules Angst (1983) *Birth Order: Its Influences on Personality* Springer-Verlag, Berlin, Heidelberg, and New York.

[19]Roberta Steinbacher (1981) Futuristic implications of sex preselection. in *The Custom-Made Child? Women-Centered Perspectives,* (Helen B. Holmes, Betty B. Hoskins, and Michael Gross, eds.), Humana, New Jersey.

[20]Marcia Guttentag and Paul F. Secord (1983) *Too Many Women? The Sex Ratio Question* Sage, Beverly Hills, London, and New Delhi.

[21]Amitai Etzioni (1968) Sex control, science, and society. *Science* 161, 1107–1112.

[22]The arguments in this article are further developed in the author's forthcoming book, *Gendercide: The Implications of Sex Selection* Rowman & Allanheld, New Jersey. 1985

Section III

Medical Decisionmaking Under Uncertainty

Introduction

In his essay, "Medical Decisionmaking Under Uncertainty," Professor Bernard Boxill addresses the following general question: Given that medical resources in our society are limited, how are we to decide whether to aggressively treat premature infants when prognosis is uncertain? In order to answer this question, Boxill examines three specific cases in which aggressive treatment for premature infants is being considered: (1) Given that all premature infants who need treatment cannot be treated because of a lack of medical resources, how are we to decide who should be treated when the decision involves a group of infants, all of whom have merely some chance to survive and become normal adults? (2) If we must decide between treating sick adults and premature infants with some chance of living to enjoy normal lives, whom should we treat? And finally, (3) which infants should receive treatment when the decision involves a group of premature infants, all of whom will be significantly handicapped if they survive?

In answer to (1), Boxill claims that infants who have the highest chance of survival at the lowest cost have the strongest claim to treatment. In answer to (2), Boxill first distinguishes between biological life and valuable life, and then argues that our duty to save mere biological life is outweighed by our duty to save lives with value. In practice, this means that we sometimes may decide to treat a premature infant before an adult. For example, we may treat a sick father who is the sole support of seven children before treating a premature infant, and yet choose to treat a low-birthweight infant before expending resources to save a man who is ninety years old and senile. Finally, Boxill argues that the principle that valuable life ought to be saved before biological life can be extended to answer question (3). When we have a group of premature infants, all of whom will be significantly handicapped if they survive, and we cannot treat all, we should first treat those whom we judge would find their lives valuable. In order to determine which infants would find their lives valuable, we must use two criteria: (a) testimony of adults having handicaps, and (b) evidence we obtain by imagining ourselves with handicaps of various sorts.

Boxill admits that these criteria do not provide perfect guides for decision-making, but nevertheless argues that they do provide some rational grounds for determining which handicapped infants should be treated when medical resources cannot be allocated to all.

Since 1958, pituitary dwarfism has standardly been treated by injections of a biosynthetic growth hormone (hGH) obtained from the pituitaries of human cadavers. Until recently, only children with growth hormone deficiencies were given hGH because supply of the hormone was limited by its source. Now, however, recombinant DNA technology allows us to produce hGH in virtually unlimited quantities, and as supply of the hormone has increased, so too has demand. With increasing frequency, parents of children who are normal, but merely short for their age, are requesting that their children be given hGH. These parents hope that their children will grow faster and become taller than they would without hGH treatment. At present, however, it is not known whether hGH will produce these effects in normal children. Indeed, very little is known about either the short- or long-term physical, psychological, and social effects of treating normal children with hGH. Thus, the problem is this: Should physicians accede to parents' requests and treat normal children with hGH, or should they refuse such requests?

In his essay, "Sharing Uncertainty: The Case of Biosynthetic Growth Hormone," Professor Martin Benjamin examines the above problem in some detail. He considers a variety of reasons that parents of normal children would appeal to when requesting that their offspring be treated with hGH, and concludes that it need not be mere vanity that prompts such requests. At the same time, he is troubled by the possible adverse effects of treating normal children with hGH. Many of the uncertainties concerning the efficacy and possible side effects of hGH treatment in normal children might be resolved by clinical research, however, Benjamin notes that research of this sort is ethically questionable. Subjects in such experiments would be placed at greater than minimal risk; research of this sort is justified only if it presents the prospect of direct benefit to subjects, and at present, we do not know that hGH injections would benefit children who are not suffering from pituitary dwarfism. In the end, Benjamin concludes that it would be "professionally irresponsible" for physicians to treat normal children with hGH whenever parents want such treatment. At the same time, he also insists that it would be wrong for physicians to reject such requests merely as a sign of parental vanity, for there are strong indications that being tall secures certain advantages in our society. The proper response, Benjamin feels, is for physicians to discuss hGH treatment thoroughly with parents, ''and to share, as fully and clearly as possible, our knowledge of the risks, costs, and uncertainty about growth hormone treatment.'' Furthermore, physicians should make parents who are legitimately concerned with ''heightism'' in our society aware that unrestricted hGH use is not the only possible response to discrimination against short men and women. There is strong evidence that parental support can be effective in overcoming many of the disadvantages of being at the lower extreme of the height curve, and where heightism is institutionalized, social efforts can be mounted to reveal and combat it.

Medical Decisionmaking Under Uncertainty

Bernard Boxill

Should aggressive treatment of premature or low-birth-weight infants be undertaken when prognosis is uncertain? Doctors know how to treat many premature or low-birth-weight infants so that they grow up to lead normal lives. But there are some premature or low-birth-weight infants that doctors do not know how to treat. Given the current state of medical knowledge, these infants cannot be saved, and their premature death is inevitable. The question posed does not ask about either of these two classes of infants. In their case, prognosis is not uncertain. Within the uncertainty that the ineluctably attends all empirical knowledge, we know what will happen to them, both if they are, or are not, treated. Nor do they present any peculiar conceptual problem. For example, the only difference between infants in the first class and normal infants, is that caring for the former costs more. Similarly, infants in the second class also present no peculiar problem. If an infant is dying and doctors know only how to prolong its life by a few hours or days, treatment will be either experimental or given to reduce the pain. These are exactly the alternatives we face when it is an adult who is dying.

I therefore set aside, in the ensuing discussion, consideration of these two classes of infants. My topic concerns only that class of infants about which doctors do not know enough to save, but know enough to give some chance of survival. This third class, of course, shades into the other two. Much experimental treatment has some chance of success, and all treatment, no matter how standard, is experimental in the sense that it offers opportunities for greater understanding that can be used to improve treatment. Yet, there are intermediate cases and it is on these that I focus.

Among such cases will be some that, if the infant survives, it will grow up to be a normal adult. The uncertainty involved in such cases does not make them radically different from the case of the infant that doctors know how to save, but only at great cost, nor, indeed, from the case of the normal infant. In each of these cases, if the infant lives it will grow up to be a normal adult. The differences lie only in costs and chances, and these shade into each other. At one end of the spectrum is the normal infant. In the middle is the premature or low-birth-weight infant that aggressive treatment will almost certainly save. At the other end of the spectrum is the premature or low-birth-weight infant that aggressive treatment may possibly save. The costs of giving the first case a high chance of survival are relatively low; in the second case they are relatively high; and in the third case, the costs of giving a low chance of survival are relatively high. One simple principle suffices to place the claims of these three infants in descending order of stringency. That principle is that where resources are limited one should place one's bets on the sure winner before the longshot—at least where the payoffs are of equal value. Given that, as the case at hand, the payoffs must be assumed to be of equal value because the lives of the three infants must be assumed to be of equal value, it follows that the first infant has the strongest case to be cared for, the second has a somewhat weaker case, and the third the weakest case.

It may be objected to that, although the principle cited is acceptable in most contexts, it is unacceptable, and indeed offensive, where what is at stake is human life. One should not, it may be protested, gamble with human life; one should seize every chance to save a human life, whatever the cost; an infant's life is a human life; consequently, as long as aggressive treatment can give an infant some chance of survival, the infant should get that treatment, whatever the costs.

In answering this objection, note first that I did not say, nor imply, that an infant that has a chance of survival only with aggressive, expensive treatment, should not get it. I said only that such an infant has a weaker claim for care and treatment than infants with better chances of survival with less expensive treatment. This position is combatible with a policy of trying to save all infants that have a chance of survival. It would require denying care and treatment to infants with a chance of survival, even with aggressive treatment, only when not all infants can be cared for and treated. Of course, because funds for caring for and treating infants are limited, on a policy level, such a situation is always with us. We cannot care for and treat all infants. Some policy for choosing which infants are treated is essential. And surely, the policy that derives from the position suggested is the one most compatible with the view that human life is valuable. If we can only try to save one of two infants, it is better that we try to save the one to which we can give a better chance of survival. To let them both die, or to try to save the one with the smaller chance to live, would be incompatible with the view that human life is valuable, or with the view that the prospective lives of the infants must be assumed to be of equal value. But the first of these views is the central idea of the objection, and the second is eminently reasonable. Whether or not one of the lives of the infant is more valuable than the other, we have no way of telling. Similarly, if resources are limited, it is better to use them to save infants that are cheaper to save because in this way we can save more infants. As in the previous case, an alternative policy would be incompatible with the view that human life is valuable, or with the view that the prospective lives of infants must be assumed to be of equal value.

Infants' claims to the care and treatment necessary for saving their lives may conflict not only with similar claims of other infants, but with similar claims of other, more mature, human beings. That is, it may happen that not all infants and other humans can be given the care and treatment they need to survive. This is a broader problem than the previous one. Yet it is also pressing. Resources are limited not only for the treatment of infants, but for the treatment of all humans. On what principles should it be shared between infants and others?

It may seem that a correct policy here should parallel that for the case of a conflict between infants' claims: That is, make no discrimination between infants and other humans, and save either the one with the better chance of survival, or if these chances are equal, the one whose care and treatment is cheaper. But there is a complication here that stems from the importance accorded to personhood in recent philosophic discussions of life and death.[1] According to these discussions, only persons have a serious moral right to life, the argument for which can be summarized as follows: To have a right to something, a being must have the capacity, at some time, to desire that thing. The right to life is the right to future existence as a subject of experiences. Therefore, to have a right to life, a being must have the capacity, at some time, to desire its future existence as a subject of experiences. But to desire something, a being must have the capacity to conceive of it. Consequently, to have a right to life, a being must have the capacity to conceive of itself as a continuing subject of experiences, and since only persons—by definition—have this capacity, only persons have serious moral rights to life. If this is accepted, it seems that a correct policy must always discriminate in favor of more mature human beings over infants. For, very clearly, the infant is not a person as we have defined personhood, and though we are assuming that it will develop normally if it survives, and thus, that it may become a person, the person it may become is, at the moment, only a possible person. The more mature human being, on the other hand, especially if it is at least a few years old, is likely to be an actual person. Consequently, if our policy is to save the infant instead of the more mature human being, it seems that, at worst, we unreasonably favor the life of a being with no right to life over the life of a being with a right to life; and, at best, that we, again unreasonably, favor the merely possible rights of a merely possible being over the actual rights of an actual being. On the other hand, a reverse policy would meet with no parallel difficulty. For if we save the person and let the infant die, the person the infant would have become had he lived remains merely possible. It never becomes an actual person with actual rights that can genuinely conflict with those of the present, actual person.

Does it follow then, that in cases of conflict, persons should always be saved before infants? Since resources are always limited, and consequently, on a policy level, we are always faced with a conflict, answering this question in the affirmative would mean alloting most medical resources to the care and treatment of persons, included aged and very sick adults with only a slim chance of survival, and almost nothing to the care and treatment of infants. Surely this is unacceptable. Yet, if the last argument of the previous paragraph is sound, such a policy seems justifiable.

The view that it is justifiable rests either on the assumption that only persons have rights to life, or on the assumption that the duty correlated to the right to life outweighs all other duties. Denying either assumption could provide the basis for a more reasonable policy. If infants have rights to life, although they are not persons, the basis for a policy of always favoring more mature humans over infants would disappear, and there would be room for a more equitable policy. Alternatively, even if only persons have rights to life, if the duty correlated to that right can be outweighed by other duties, there would be room for a policy alloting a generous portion of medical resources to the care and treatment of infants.

But although the assumption that infants have rights to life makes room for a policy that allots resources more equitably between infants and adults, it does not, by itself, support a policy that is altogether reasonable. If we assume that infants have rights to life, but do not assume that the duty correlated to the right to life can be outweighed by other duties, we could justify a policy that required that the resources of a society be allocated to the saving of the lives of such infants and other humans up to the point at which greater allocation would endanger the lives of healthy humans. Such a policy would severely reduce the quality of life in the society. But, is it for this reason unacceptable? That is, does the right to a certain quality of life outweigh the right to life itself?

To have a right to something is to have a very strong claim on others to either provide us with, or at least not interfere with, our possession of that thing. Clearly then, the point of our having rights at all is that they secure to us things that are particularly valuable to us. We have rights to liberty and happiness, and rights to vote and to stand for public office because these things are very valuable to us. In particular, we have rights to life because life is very valuable to us. But life is valuable to us because it is a condition for our experiencing or being aware of all other things that are valuable to us. If pleasure, beauty, achievement, and love are valuable to us, we must be alive to be aware of beauty, to achieve, to love, and to feel pleasure. But it does not follow that life itself, because it is in the above senses a condition of value, is valuable. Although life is a condition of beauty, achievement, love, and pleasure, it does not follow that mere biological life, without beauty, achievement, love, or pleasure is valuable. Indeed, it seems clear to me that mere biological life, the life, for example, of one in an irreversible coma, has either no, or little, value in itself. If a universe with such a life is better than a universe with no life at all, it is only barely so. Consequently, given that rights secure to us things that are valuable to us, then if there is a right at all to mere biological life, and a correlate duty to preserve that right, that right and that duty are vastly outweighed by the right to a valuable life and the duty to preserve such a life. In other words, the duty correlated to the right to life can be, and indeed often is, outweighed by other duties, in particular, the duty correlated to the right to valuable life. But the policy of preserving life without regard to its quality gives an absolute weight to bare life relative to valuable life, and is for this reason, unacceptable.

Returning to the view that maturer persons should always be saved before infants, we can now explain why it is false. In the first place, the claim that only persons can have rights to life is controversial. The crucial premises of that claim were that only persons can desire life, and that to have a right to something one must have the capacity to desire it. But, although the first of these premises is unarguably true, we can now see that the second is arguably false. Given the nature of a right as a very strong claim on others, either that they provide us with, or at least not interfere with, our possession of something, then, as I suggested, it seems clear that the point of having rights is to secure to us things that are particularly valuable to us. An inspection of the nature of a right does not similarly suggest that to have a right to something we must, at some time, have the capacity to desire it. Those who maintain that only persons can have rights to life must prove this crucial claim because it is not obvious. It does not, for example, follow from my suggestion. Things can be valuable to us even if we lack the capacity to desire them. In particular, life can be valuable to the infant even if it lacks the capacity to desire life. Consequently, until my suggestion is shown to be false, we have no reason to say that infants have no rights to life because they lack the capacity to desire life.

Furthermore, since, as I have argued, the duty to preserve life is outweighed by the duty to preserve valuable life, it does not follow that we must always save the person before the infant. It depends on which life is more valuable. If the person is at the end of his life and treatment can only prolong it for a few months, or if his life will be burdened by a great handicap, whereas the infant has prospects of a normal life, then the infant should be saved before the person.

The uncertainty I have so far considered is that of the survival of premature and low-birth-weight infants. In particular, I have assumed that if these infants survive, they will grow up to be normal. But there is a further uncertainty to consider. When, even with aggressive treatment, prognosis of premature or low-birth-weight children is uncertain, often, part of the uncertainty is that although the infant's life may be saved, the infant may grow up with a significant handicap. The problem this presents is different from the last one. It is one thing to go to extraordinary lengths and expense to give an infant a chance to grow up to be a normal adult, with a normal adult's opportunity for happiness. It is another thing to go to extraordinary lengths to give an infant a chance to grow up significantly handicapped, with perhaps significantly reduced opportunities for happiness. What should be done in this sort of case?

A somewhat similar question can arise for aggressive treatment of very sick adults. Sometimes such treatment may not save their lives, and if it does, it may not save them from significant handicap. Even when a policy designed to save valuable lives ensures that resources for treatment are available, it is reasonable to let the adult decide for himself or herself whether to accept treatment. The question is not whether one's life will be, in some objective sense, valuable. If the policy is well designed, we can assume that it will be. The question is what one thinks. If an adult refuses to give her or his consent to treatment, that is a very good, if perhaps not decisive, reason not to give them treatment; if, on the other hand, she or he gives her or his consent to treatment, that is also a very good, although again not decisive, reason to given them treatment. But, consent, which features so prominently in the adult case, does not even arise in the infant case. Manifestly, the infant can neither give, nor refuse, consent to treatment.

Of course, dangerously ill adults are often somewhat like infants in that they can neither give, nor refuse, consent to treatment. Yet, in this case too, the idea of consent can be used in a way that it cannot be used in the case of the infant. For example, although the comatose adult can neither give, nor refuse, consent to treatment, we can perhaps say truly of her or him, although not of the infant, that he or she would consent to treatment, or would not consent to treatment. The comatose adult has a determinate enough personality, values, and attitudes toward risk-taking, such that if she or he were conscious and aware she or he either would or would not consent to treatment. Because of this we can use the idea of consent to decide whether or not to give aggressive treatment to comatose adults. If the doctor can determine what sort of person the patient is, and in particular, what the patient's goals, plans, and attitudes to risk-taking are, that doctor can make an educated guess at whether or not the patient would consent to aggressive treatment. But clearly, a doctor faced with the question of whether to undertake aggressive treatment of a low-birth-weight or premature infant, cannot even begin to try to determine what sort of person it is, nor what its plans, goals, or attitudes to risk-taking are. The infant has no goals, plans, or attitudes to risk and never had any. Indeed, as we have seen, it is not even a person.

Consent theory, therefore, can offer no help here. We must go back to the principle of the previous section that the duty to save valuable life outweighs the duty to save less valuable life. But applying the principle is more difficult here. There, the problem was to compare the value to an adult of living a few more months or years, perhaps in pain or handicapped, with the value to an infant of a whole life. Here, the much greater problem is to compare the value to a human being of a life handicapped in some way, to the value to another human being of a life handicapped in another way.

I do not mean to discount the difficulties of the first comparison nor to suggest that it can be undertaken lightly. It will, of course, always be saturated with uncertainty. Still, there are often assumptions we can make with a fair degree of certainty, and these can form the basis of a comparison. To the extent that we can know the personality, likes, and dislikes, and plans of the life of an adult, we can guess how profoundly a handicap will affect him or her. Although there are unknown strengths and weaknesses, resiliencies and frailties in every person, we can often hazard a guess that becoming a quadriplegic will be a greater loss to the athlete or outdoorsperson than to the sedentary; that the loss of a hand will be a greater loss to a concert pianist than to a bookkeeper; and that Alzheimer's disease will be more devastating to an intellectual than to a gardener. Similarly, although the lives of persons with normal abilities vary in value, and vary in value for them, almost all persons with normal abilities not only value their lives, but value them very greatly.

Parallel assumptions can usually not be made, or if they can be made, can be made only with even greater uncertainty when the comparison is between the lives of infants who will grow up with different kinds of handicaps. Is it possible, or even meaningful, to compare the lives of the severely retarded with the lives of the severely crippled? I do not deny that the comparison is difficult, nor do I even affirm that it is possible. But I do argue that a coherent policy is possible.

The difficulty suggested in the above paragraph stemmed from the fact that there are radically different kinds of handicaps. Yet, it is this difference that, I think, permits the formulation of a coherent policy. Let us divide handicaps into three classes, those involving pain, those involving physical disability, and those involving mental disability. Infants whose handicaps fall into the third class present the gravest uncertainty. Those falling into the first or second class present a more tractable uncertainty, and I will deal with them first. I will argue that when an infant is likely to grow up mentally normal, although with the prospect of a life of continual pain or a physical disability, we have at least some firm basis for judgment. This is because of two peculiarities of normal mental ability. The first has to do with the fact that because we have normal mental abilities ourselves we can have at least a slim basis for guessing how valuable another mentally normal, but handicapped, person conceives his or her life to be. The second has to do with the fact that having normal mental abilities, a person with a life of pain or physical disability, can give his or her testimony that he or she does or does not find his or her life valuable.

Taking the first point, I do not mean to discount the difficulty of imagining what life will be like for infants growing up with severe physical handicaps. For example, unlike the case of a normal person who becomes physically handicapped, there cannot here be any basis in the personality, likes, dislikes, and plans of the infants to guess at how they will respond to their handicaps. We are faced, not with handicaps befalling previously normal persons with already formed personalities, but with handicapped infants growing into handicapped persons and having their personalities formed by their peculiar experiences. Accordingly, it is far more difficult for us to enter imaginatively into the life of the persons such infants will become, and thus, far more difficult to guess at how valuable they will find their lives. Comparing it with the life of a person who becomes paralyzed, or using the testimony of such a person, is not likely to be very fruitful. The person paralyzed from birth, for example, has an advantage over one who becomes paralyzed because he or she never got attached to running or dancing and consequently feels no loss at his or her disability. Also, given that one would presumably develop likes appropriate to one's condition from the start, one would not, as does the person who becomes paralyzed, have to go through the agonizing process of reeducating and reconditioning oneself to like what he or she can do and have. But, he or she also has disadvantages. Even if one never gets attached to running, he or she can still want to run and can deeply resent his or her incapacity. And it is not clear that it is better never to have run and never to miss it, than it is to have run and to miss it. Still, despite these difficulties, it is possible, I think, for us to have some idea of whether an infant with normal mental ability, but with a severe physical handicap, will grow up to find his or her life valuable. All of us are born with some handicap, however minor, and with an effort of imagination can use it as a basis for guessing the quality of life of others with more serious handicaps.

Taking the second point, I am not proposing that we must wait for the infant to grow up and report whether his or her life is valuable. This would require that all low-birth-weight and premature infants be given aggressive treatment and I am assuming that resources are too limited for that. I am proposing that, insofar as we can have the testimony of others who grew up with similar handicaps, we have some basis for saying that the infant will or will not have a valuable life. The two points are meant to act as a check on each other. We must try to imagine what life would be like blind, crippled, or in pain, but we must compare our findings with the testimony of those who have lived these lives. On the other hand, this testimony can not be taken as final. A quadriplegic may say that his or her life is not worth living, but we may be able to see that this verdict stems from alterable circumstances other than his disability.

Given that there is a strong duty to save valuable life, and a much weaker duty to save life that is less valuable, these two points support the following: (1) When, as a result of an imaginative effort at identification with the prospective life of an infant, checked by the testimony of those who have lived with the handicap he or she is likely to have, we judge that the infant will find his or her life valuable and will desire to live, then that infant has, in the present context, a strong claim to aggressive treatment, and probably should receive such treatment. (2) When, on the same grounds, we judge that an infant will not find his or her life valuable, will desire not to live, and will wish that he or she has never been born, then that infant has, in the present context, a weak claim to aggressive treatment and probably should not receive such treatment.

Furthermore, I think that the following can also be established: Infants whom we judge will find their lives valuable, and who will desire to live, have almost always the strongest claim to aggressive treatment. Correspondingly, infants whom we judge will not find their lives valuable, will desire not to live, and wish that they had never been born, have almost always the weakest claim to aggressive treatment. In practice this means that infants whose lives will be marred only by physical disability or pain should be treated before infants who will grow up mentally disabled, unless the physical disability or pain is likely to be extreme.

Because infants who grow up to have the mental capacity to desire or not to desire life are persons, and those who grow up without that capacity are not persons, it may be objected that I seem here to give precisely that weight to personhood that I denied it in the previous section. It should be first noted that the policy I recommend does not invariably prefer persons over nonpersons. On the contrary, it recommends that nonpersons who lack the capacity to desire or not to desire the lives they have, but whose lives are nevertheless pleasant, enjoyable, and thus, valuable to them, be preferred over persons who do not desire the lives they have. But, that point aside, it remains that, at least where there is a desire for life, the policy of the present section almost always prefers the person over the nonperson, and this may seem inconsistent with the policy of the last section that preferred infants with the prospect of valuable lives—and infants are, by assumption, not persons—over adult persons with the prospect of far less valuable lives.

My position in the previous section was that the capacity to desire something was not a necessary condition for having a right to that thing, but rather that the capacity to benefit from something was a necessary condition for having a right to that thing. On the basis of that position I concluded that if life would be valuable to the infant, and thus, a benefit to it, it could have a right to life even if it lacked the capacity to desire life. But this conclusion is perfectly consistent with a policy of giving preference to a being who desires the good it has a right to over a being who does not, or cannot, desire the good it has a right to. On the contrary, such a policy seems required by the duty to save valuable lives over less valuable lives. A being who gets a good he or she desires enjoys a greater benefit than a being who gets a good he or she does not desire; and a being who fails to get a good he or she desires suffers a greater loss than a being who fails to get a good he or she does not desire. For, in the first case, in addition to the enjoyment of the good, there is the satisfaction of desire. Consequently, assuming lives that are otherwise equally valuable, it seems clear that the duty to save valuable lives requires that we save the infant who will become a person who desires his or her life over the infant who will never become a person and who, accordingly, will never desire his or her life.

It may be objected that even if this argument is sound, it only shows that where lives are otherwise equally valuable, the life of the person who desires his or her life is more valuable than the life of the individual who lacks the capacity to desire his or her life. It does not show that, where lives are otherwise unequally valuable, the life of the person who desires his or her life is more valuable than the life of the individual who lacks the capacity to desire his or her life. Suppose, for example, that the life of the former is full of pain and that he or she barely desires it, whereas the life of the latter is full of simple joys and pleasures, although he or she lacks the capacity to desire it. Surely, my argument does not show that the life of the former is more valuable.

My answer to this objection is that it does not give sufficient weight to the uncertainty in the judgment that the life of the individual who lacks the capacity to desire his or her life is valuable. If it is difficult to guess at what another mentally normal person's life is like, it is far more difficult to guess at what a mentally abnormal person's life is like. We cannot use even the simplest joys and pleasures, which we seem to have in common with the mentally subnormal, as a basis for judgment. The value of these joys and pleasures to us is deeply enhanced by, and thus depends on, our perception of them, and this, in turn, depends on mental ability. I admit, of course, that I am dealing here with a matter of degrees. We can guess at the life of the only slightly retarded, and often such a life may be deemed more valuable than the life of a mentally normal person who is severely disabled physically and suffers constant pain. But such borderline cases will be in the minority. On the reasonable ground that a certain good should be saved before an uncertain good, the deeper the mental retardation, the more should preference be given to the mentally normal.[2]

My assumption in the foregoing is that resources are not so plentiful that we can perform both our duty to preserve life and our duty to preserve valuable life. Consequently, hard choices have to be made, often under uncertainty. This uncertainty is especially great when we are dealing with premature and low-birth-weight infants. Often we are uncertain whether these infants will live or die even with aggressive, expensive treatment; and uncertain whether, if they live, they will develop normally. The best policy would be to try to save all such infants. Unfortunately, this is inconsistent with a policy of preserving valuable life. Some infants must, therefore, be given preference over others. My argument, in general, is that the policy most consistent with the duties to preserve life and to preserve valuable life is one that prefers those who, at relatively low cost, can be given relatively high chances of surviving and living the most valuable lives over those who, even at relatively great cost, can be given only relatively low chances of surviving and only relatively low chances of living valuable lives.

One final point must be mentioned. It is well known that premature and low-birth-weight children are most common among the poor. In most cases then, even if premature and low-birth-weight infants survive and develop normally, the poverty of their parents and environments will likely detract from the value of their lives. Should this be taken into consideration? In other words, should premature and low-birth-weight infants of the well-to-do be given preference over premature and low-birth-weight infants of the poor. Some authors suggest that this could be a justifiable policy. Though I can not argue for it here, I most emphatically do not agree.

Notes and References

[1]For example, Laura Purdy and Michael Tooley (1974) Is abortion murder? *Abortion: Pro and Con* (Robert L. Perkins, ed.), Schenkman, Massachusetts.

[2]My position here does not contradict my earlier argument that, though infants do not desire life, they can have rights to life, if life will be valuable to them. For though I argue here that desiring life is the best criterion that life is valuable, I do not say that it is a condition of life being valuable.

Sharing Uncertainty

The Case Of Biosynthetic Growth Hormone

Martin Benjamin

Introduction

"If you think the problem is bad now, just wait till you've solved it." This (or something quite like it) is called Epstein's Law. I don't know who Epstein is, but the Law bearing his (or her) name, like the more well known "laws" attributed to Murphy, Parkinson, and others, neatly captures an important truth about the human condition. Certain solutions create as many, if not more, problems as they solve.

The discovery and impending approval by the Food and Drug Administration (FDA) of biosynthetic human growth hormone (GH) for the treatment of short stature provides additional confirmation of Epstein's Law. This new source of GH is a welcome development for children whose normal growth is restricted by insufficient or faulty GH of their own. At the same time, however, the possibility of a pratically unlimited supply of GH is creating unprecedented demands that raise vexing problems and questions for parents, pediatricians, family practitioners, endocrinologists, and society at large.

Compounding these new problems is our limited knowledge and understanding. "It is the legacy of science," Jay Katz has recently observed, "that scientific activity produces not only new knowledge but also new ignorance." In what follows I will show how the discovery and manufacture of biosynthetic GH generates new possibilities that are beset with ignorance and uncertainty.

I will begin by describing the problem to which the development of biosynthetic GH is a partial, but most welcome, solution. Then I will show how, in this instance, a solution to one problem creates many others. Each of these other problems, I will suggest, is compounded by various forms of ignorance and uncertainty. After identifying the medical profession's characteristic reluctance to share uncertainty with patients, I shall make some concluding recommendations as to how these new problems might best be addressed.

Background

Pituitary Dwarfism

Approximately 10% of those children whose height is two standard deviations or more below the mean for their age suffer from a faulty or inadequate supply of GH. This hormone, produced by the pituitary gland,is vital for normal growth. Children who remain deficient in normal GH are unlikely to become taller than 4 ft. Although many have been able to adjust to their condition, the psychological effects of what is called "pituitary dwarfism" can be devastating. Adults this short are also functionally handicapped. Unable to drive a car, use drinking fountains, reach elevator buttons, and so on, they are at best inconvenienced and at worst seriously restricted with regard to normal vocational and other opportunities.

Modern treatment of pituitary dwarfism began in 1958 when the growth rate of a 17-yr-old prepubescent boy was more than quadrupled by injections of GH taken from the pituitaries of cadavers. Since then, GH from cadaver pituitaries has been the standard treatment for children with GH deficiencies. The supply of GH has, however, been limited by its source. A typical 2-yr course of treatment requires hormone from 50 to 100 pituitary glands; and in the United States, pituitaries cannot be harvested from cadavers unless specifically donated for this purpose. Because supply falls short of demand, not all who might have benefited from this treatment have received it. The National Pituitary Agency has rationed the limited American supply by restricting its use to very short children who are demonstrablty deficient in their own GH. The quantities allotted to each patient have been limited, and treatment, if successful, has usually been discontinued when boys reach 5 ft. 6 in. and girls reach about 5 ft. 4 in.

All this is now about to change. Among the first clinical applications of recombinant DNA technology is the development and manufacture of biosynthetic GH. Growth hormone practically identical to that produced by the normal pituitary gland can now be produced in unlimited quantities. The supply will, from now on, be limited only by the demands of the marketplace. For the first time the availability of cadaver pituitaries will no longer be the principal factor in deciding whether a child can be given GH and if so, how much. This is indeed good news for hypopituitary children and their parents. But once approved by the FDA, it is unlikely that treatment with GH will be restricted to children suffering from GH deficiencies.

New Demands

Pediatricians, family physicians, and endocrinologists frequently see parents who are concerned about their children's height. In many cases these children are otherwise normal, but shorter than others their age either because of a familial pattern of delayed growth or because their parents are also comparatively short. Children falling into the first category are shorter than their peers throughout childhood, particularly during adolescence. Bone age (the maturational index of the bones of the hand and wrist) is retarded, and often there is a family history of delayed puberty and growth. Although this may be a cause of embarrassment and distress, the condition is temporary. Growth, though slower than in other children, proceeds for a longer time and the child's final height, like that of his or her parents, is usually normal. The customary treatment is the physician's sympathetic assurance to the child and parents that things are likely to turn out fine. Children falling into the second category are, and are likely to remain, short, although their final height is entirely normal given their family background. Height is one of the most heritable traits and the perfectly healthy child of short parents is, for genetic reasons, also likely to be short.

Short stature attributable either to what is called "constitutional delay" or genetic endowment is not a function of GH deficiency. Yet anticipating the availability of biosynthetic GH, some parents are already requesting that these short, but otherwise normal, children be treated. Because excessive GH from a pituitary tumor can make a giant out of a normal child, there is some reason to believe that additional GH could have a similar, but more controlled and limited, effect in children who make their own GH.

Moreover, because desired height is a relative matter, requests for treatment may also come from parents of medium, or taller than average, height. For example, a mother who wanted to be a model, but whose aspirations were thwarted by being 5 ft 5 in. may want her daughter to have the opportunity she was denied. A 5 ft 9 in. father who wanted to be a college basketball player, but who, though highly skilled, was not tall enough to compete at that level, may want to spare his athletically inclined son or daughter a similar fate. Although their children are likely to be of average height, they may also seek to have them treated with biosynthetic GH.

The Height of Success

In an editorial (''The Height of Vanity''), the *New York Times* anticipated the new demands created by a limitless supply of GH and recommended that treatment be restricted to children whose pituitaries produce too little effective hormone.[2] Other uses were labeled ''cosmetic'' and, as the title suggests, said to be motivated by vanity. But the concerns of many parents, I suspect, go deeper than this. Parents of otherwise normal children whose growth is either constitutionally delayed or genetically limited may want to keep them from being bullied or ridiculed by their peers. They may be afraid, too, that their children will become overly passive and timid, or else compensate by developing a ''Napoleon complex'' and constantly strive for power and control over others.[3] In addition, they may believe, not unreasonably, that other things being equal, taller people are more likely to be successful than shorter people, provided they are not excessively tall.

A growing body of evidence suggests that height does affect job prospects, salary, and chances of winning political office.[3] In one study, for example, 140 recruiters were asked to choose between equally qualified hypothetical job candidates, although one was described as being 6 ft 1 in. and the other was characterized as being 5 ft 5 in. Seventy-two percent preferred the taller candidate, who thus had nearly a three-to-one advantage over the shorter. One recruiter selected the shorter candidate and only 27% acknowledged that the two were equally qualified.

Other studies suggest that starting salaries of college graduates are determined in part by height. One survey compared the salary differences between male library science graduates who ranged between 6 ft 1 in. and 6 ft 3 in. and those under 6 ft with the salary differences between those who were in the top half of their class academically and those in the lower half. The average difference in starting salary between the taller and the shorter graduates was more than three times greater than the difference between the more and less academically qualified.

Height also appears to affect promotions and raises. Using a sample of 5085 men who had passed an Air Force cadet qualifying exam, one study showed that after 25 yr, those who were 5 ft 6 in.–5 ft 7 in. earned $2500 less per year than those who were 6 ft–6 ft 1 in.

Since 1904, the taller candidate for President of the United States has been the victor of 80% of the elections, whereas Democrats or Republicans can claim only a 50% success rate. Of all the Presidents, only James Madison and Benjamin Harrison were shorter than average for American males at the time they were elected.

The results of these and other studies, as well as anecdotal or personal experience, may prompt parents of short, or even somewhat taller than average, children to consider a course of medical treatment that may make their offspring taller. They are motivated by more than simple vanity and they regard the value of increased height as more than merely cosmetic. Upwardly mobile, middle-class parents preoccupied with their children's health, education, the quality of their neighborhood, and so on may be especially concerned about their children's height. A trial lawyer, for example, who feels strongly—and with some justification—that being a few inches taller would significantly increase his effectiveness before a jury, his chances for a judgeship, or his political success, may quite understandably be interested in a medical intervention that promises to make his children a bit taller.

Risks, Costs, And Uncertainty

Yet the extended use of GH is fraught with risks, costs, and uncertainty. We know comparatively little about the long- and short-run effects of administering GH to children who appear to have adequate levels of normal GH. Nor do we know enough about the psychological and social effects of this new form of treatment. Until parents and physicians are more aware of the risks, costs, and uncertainties of administering GH to children who do not appear to be GH deficient, they are unprepared to make an informed judgment about its use.

Diagnosis

Pediatricians can make fairly (though not perfectly) accurate predictions about a healthy child's final height.[4] Family history, an accurate record of growth rate based on at least 6 mo observation, and X-rays of the hand to determine bone age as compared with chronological age are all useful for this purpose. Predictions based on these factors are generally accurate within a range of a few in. But they are not infallible. Growth is an extremely complex, multifactorial process that is not wholly understood, and physicians' informed predictions of children's final height are occasionally off the mark. Thus, a decision to embark on a course of treatment with GH, for an otherwise normal, healthy child cannot be based on knowledge that he or she is otherwise certain to fall below a desired final height. Assuming that the physician's measurements have been accurate, it is at best likely that the child will, without treatment, end up within the predicted range.

Treatment

Although no market price has been set for biosynthetic GH, it is unlikely to be less expensive than pituitary GH which presently ranges from $6000–9000/treatment yr. Biosynthetic GH is currently being promoted to pediatricians by the manufacturer. It would be suprising if the developers were not to seek the highest return on their investment that the market would allow.

Treatment with GH is likely to be most effective with children between the ages of 7 and 14. To date, the response to GH has usually been best during the first year, with efficacy falling off during subsequent years. A typical course of treatment will involve intramuscular injections of 1–2 cc 3x/wk. The injections—administered like insulin by the parents—will be painful and must continue anywhere from one to several years. In most cases the minimum course of treatment will be 1–2 yr.

Of three things, then, we may be sure. Treatment with GH will be costly, long, and not particularly pleasant. The overall effect on the individual child and possible long-term social consequences are, however, much less certain.

Effects On Growth

Though the desired effect of treatment is an increase in the child's final height, we are not yet in a position to say whether, or which, apparently normal children will show such an increase. The few studies that have been done to date suggest that a significant number of a comparatively small set of healthy, short children given hGH for periods of 6 mo experienced an increase in growth rate. In one study, 6 of 14 prepubertal, short, normal children experienced such an increase whereas in another, the number was 8 of 10.[6] As Louis E. Underwood recently concluded, "Although results vary from one study to another and the collective experience is small, it appears that there is a better-than-average chance that GH treatment will accelerate growth in short children in whom thorough evaluation uncovers no etiologic factor."[7] At the same time, Underwood acknowledges that these results raise as many questions as they seem to answer.

For example, an initial increase in growth velocity may not result in a significant increase in final height. It will, at best, be many years before we know whether GH treatment in normal children actually makes them taller as adults or whether the final height they would have attained without treatment will simply be reached at an earlier age. Moreover, since growth is an agonizingly slow process, considerable time, discomfort, drug exposure, and cost are necessary before we can retrospectively determine the treatment's efficacy in a particular child. Thus, we will want an accurate way to distinguish those whose eventual height is likely to be raised by treatment from those who are not likely to respond. "How do we identify those children who will benefit from GH therapy from those who will not?" Underwood asks. Then, answering his own question, he says, "Using the tools now available, children who will respond cannot reasonably be distinguished from those who will not."[7]

Side Effects

Apart from costs, pain, duration, and uncertainty of results, another drawback to GH treatment is the possibility of adverse side effects. Administering GH to children who are not GH deficient may produce diabetes mellitus, hypertension, atherosclerotic coronary artery disease, or cerebrovascular disease. A small percentage of patients may form antibodies against GH.[8] In addition, impurities from the *Escherichia coli* bacteria that "make" the GH may cause side effects. Like many medical interventions, then, the comparatively small but significant possibility of unwanted effects complicates the weighing of risks and benefits.

Psychological Effects

Less obvious, but no less significant, are the psychological effects on the child if treatment falls short of expectations. Even if certain children were not particularly self-conscious or worried about their height before treatment, a long series of painful injections may convince them that they were suffering from a serious disability. What else, they might ask, could justify having to undergo such an ordeal? Thus, the treatment may plant the seeds of doubt and inferiority, when with parental acceptance and encouragement they might never have existed.

According to a study by Diane Rotnem and her colleagues of the psychological effects of GH treatment in 11 GH-deficient children, "Expectations of growth were heightened by elaborate endocrine evaluations and hospital admissions. The unrealized expectations contributed to the child's feeling that he was unacceptable as he was." Further, as growth fell below parental expectations, the children "became acutely aware of the increased parental anxiety and disappointment, experiencing some of their parents' moods in themselves . . . . This mirroring was apparent as the children directed anger or self-deprecatory comments toward themselves."[9] We should be quite concerned, then, about the psychological impact on otherwise normal children if the treatment is ineffective or less effective than expected. In some cases our intervention is likely to create the very problems we seek to avoid.

Social Effects

Finally, even if all of our uncertainty about the effects of administering GH to particular children can be resolved and we can, if we wish, raise the height of individual children with few ill effects, problems of justice and public policy remain when we view the matter from the perspective of society as a whole. At present cost, neither Medicaid nor most private insurers are likely to pay for such treatment. Who, then, will be the recipients? Mainly, children of upper, upper-middle, and upwardly mobile middle-class parents who can afford to pay for what they regard as every possible advantage for their youngsters. Thus, if height is in fact correlated with success in certain occupations and pursuits, present differences among social and economic classes will become more deeply entrenched. Although we cannot be certain, the children of the wealthy could, with the aid of this medical intervention, increase their general advantage over the children of the poor.

Further Research

Many of the doubts about the efficacy and possible side effects of GH treatment in short and not-so-short, otherwise normal children might be resolved by scientifically well-designed clinical research. Yet clinical research in this area is ethically questionable.

Ethical Considerations

The difficulty concerns the risks involved in administering GH to children who do not appear to be GH-deficient. According to the recently revised federal regulations governing research on children, there is a close connection between the degree of risk that a child is permitted to undergo and the prospect that the research will provide him or her with a direct benefit.[10] If children are subjected to only "minimal risk," then the regulations require only the informed "assent" of the child and the informed "permission" of the parent(s) or guardian. "Minimal risk," as defined in the regulations, means that "the risks of harm anticipated in the proposed research are not greater, considering probability and magnitude, than those ordinarily encountered in daily life or during the performance of routine physical or psychological examinations or tests." By this definition the risks of receiving GH are greater than minimal. Apart from possible biological side effects, the treatment may, as noted above, have untoward psychological effects. Some of the GH-deficient children observed by Rotnem, for example, "saw themselves as victims of painful and useless injections, and persistently or intermittently resisted shots and directed angry feelings toward their parents."[9]

If the risk is greater than "minimal," the research can be justified only if it presents the prospect of direct benefit to the individual subjects. The easiest studies to justify will therefore be those involving children with GH deficiencies or children without such deficiencies who are nonetheless likely to be so short as adults that they will be functionally handicapped in our society. But whether an increase in height for healthy children whose predicted adult height is simply below the 10th percentile is enough of a benefit, whether it is a "direct" benefit, and whether it is a benefit at all, are difficult to determine. And the taller the child is otherwise likely to be, the weaker the claim that the therapy promises to be of "direct benefit."

A Difficult Dilemma

Those who believe that further research will reduce or eliminate the uncertainty currently surrounding decisions about the use of GH encounter a difficult dilemma. The most justifiable studies from an ethical point of view will be those that are restricted to children who are either GH-deficient or who, though otherwise normal, are predicted to be execptionally short. These studies, however, will leave unanswered questions about the efficacy and risks of GH therapy for taller children. Yet studies on these children are ethically suspect because the greater than minimal risk to the subjects cannot be offset by the promise of direct benefit. To the extent that further research is ethically justified, then, it will not resolve our questions about the extended use of biosynthetic GH; and to the extent that certain studies may resolve these questions, their use of children as subjects will be ethically questionable.

Coping With Uncertainty

"The denial of uncertainty, the proclivity to substitute certainty for uncertainty, is," as Jay Katz has observed, "one of the most remarkable human traits. It is both adaptive and maladaptive, and therefore both guides and misguides."[1] Doctors are not less human than anyone else in this regard, and in his recent book, Katz analyzes the way physicians cope with uncertainty. Although I cannot convey here the depth and complexity of his complete account, I want to focus on one aspect that is especially relevant to decisions about the use of biosynthetic GH.

Acknowledging And Denying Uncertainty

Physicians, Katz maintains, respond to uncertainty in complex ways: "They will acknowledge medicine's uncertainty once its presence is forced into conscious awareness, yet at the same time will continue to conduct their practices as if uncertainty did not exist (p. 166)." This is reflected in a significant gap between the extent to which physicians will acknowledge uncertainty when talking with each other and the extent to which they will acknowledge it when talking with outsiders. "[T]he reality of medical uncertainty," Katz adds, "is generally brushed aside as doctors move from its theoretical contemplation to its clinical application in therapy and, even more so, in talking with their patients." He recounts a conversation with a surgeon-friend in which they readily agreed on the degree of medical ignorance and uncertainty surrounding the question of the best treatment for breast cancer. Then, in response to a question about what he would tell a patient, the surgeon spoke of a recent experience with a particular patient. A few days earlier he had been consulted by a woman who had just been diagnosed as having cancer of the breast. At the outset of their discussions, he told Katz, he briefly mentioned a number of treatment alternatives without indicating that any of them deserved serious consideration. He then proceeded to impress upon her the need to submit immediately to a radical mastectomy. Katz writes,

> I commented that he had given short shrift to other treatment approaches even though a few minutes earlier he had agreed with me that we still are so ignorant about which treatment is best. He seemed startled by my

comment but responded with little hesitation that ours had been a theoretical discussion of little relevance to practice (p. 167).

According to Katz, this conversation illustrates that the uncertainty of medical knowledge is often not at issue. Physicians will, at least among themselves, usually acknowledge that medicine is full of uncertainty. What is at issue, however, is the extent to which this uncertainty should be shared with patients. In general, Katz maintains, physicians believe that patients should not be fully apprised of the limits of medical knowledge and the extent of medical ignorance.

Katz examine a number of justifications that physicians give for preserving this discrepancy and finds them wanting. Tradition, for example, emphasizes the therapeutic value of faith, hope, and reassurance, and seems to require that doctors be regarded as "bearers of certainty and good news." But, insofar as this may require dissembling and deception, he suggests, it may bring about the very result it seeks to prevent. Patients may become alienated from, and distrustful of, their physicians when misrepresentations and falsehoods begin to unravel, as they usually will. Intimacy is compromised and with it the therapeutic value of the relationship itself. Thus, contrary to the conventional wisdom, "it may turn out that an acknowledgement of uncertainty will enhance physicians' therapeutic effectiveness, because it demonstrates honesty in the face of uncertainty and a willingness to be more engaged with their patients than is possible when communications are beset by evasions, half-truths, and even lies(p. 193)."

Another argument physicians give against acknowledging uncertainty is that candor about the limitations of medical knowledge may indirectly harm patients by driving them into the arms of quacks who will promise even more. Although admitting that some patients with an exceptionally low tolerance for uncertainty and a deep need to be assured and cared for by "miracle workers," may succumb, perhaps unfortunately, to the seductive promises of charlatans, Katz denies that the care of the vast majority of patients should be compromised for their sake. Patients in the latter category should be reassured that their physicians will be truthful about their limitations and profess only that certainty and exercise only those skills they truly possess. Moreover, he adds, "Acting out of fear that any acknowledgement of medicine's limitations will drive patients into the arms of quacks has its own dangers. In promising more than medicine can deliver, physicians adopt the practices of quacks and are themselves transformed into quacks (p. 201)."

A third argument for not disclosing uncertainty turns on concerns about the economic costs of the more thoroughgoing discussions such disclosures will require. Katz admits that acknowledging uncertainty will take some time and will be reflected in physicians' fees. But he adds that "it is not all clear how much time conversation will take once doctors know what needs to be talked about, and how and why they should talk (p. 201)." Moreover, since physicians have always maintained that cost should not be an impediment to good patient care, it is surprising that all of a sudden in this context they appear to be so cost conscious. Upon analysis, Katz suggests, "it may turn out that physicians' concerns over the economic costs of conversation may also mask an underlying concern: avoidance of the uncomfortable role of being the bearer of uncertainty (p. 202)." Finally, insofar as disclosure of uncertainty persuades patients that certain costly procedures may be optional rather than necessary, overall patient costs may be reduced.

Apart from these ostensible justifications, there are, Katz recognizes, strong social and structural factors reinforcing the reluctance to disclose uncertainty. Medical education, for example, emphasizes the quest for certainty. Medical students "are led to attribute much of their uncertainty not to its inevitable presence or to its pervasiveness but to their lack of training, to not having as 'yet developed the discrimination and judgment of a skilled diagnostician' (p. 185)." Medical training and socialization, the profession's demand for conformity (which emphasizes orthodoxy and authority), and the pursuit of specialization all combine, according to Katz, to limit a physician's acknowledging uncertainty. In addition, emphasizing the differences between the certainty of doctor and patient helps maintain an imbalance of power that many physicians are reluctant to relinquish: "Physicians' power and control are maintained not only by projecting a greater sense of centainty than is warranted but also by leaving patients in a state of uncertainty, not in the sense of shared uncertainties but in the sense of keeping patients in the dark . . . Doctors' acknowledgement of their uncertainties would significantly lessen this source of patient manipulation (p. 198)."

Although he has no illusions about how quickly things may be turned around, Katz argues strongly that patients must be let in on the secret that physicians will, when they are candid, share with each other: Medicine is, and is likely to remain, fraught with uncertainty. Until physicians can admit various uncertainties to patients, the notions of informed consent and shared decision-making will remain unrealized ideals. Moreover, it is naive to believe that the aura of certainty and infallibility projected to the patient will not spill over into the physician's own thinking. "Masks," Katz points out, "can deceive not only the audience but the actor as well. The mask of infallibility makes it more difficult than it otherwise would be for physicians to explore their own doubts and uncertainties, and precludes acknowledging them to patients (p. 199)." As a result, choices that, all things considered, would be preferred by both patient and physician, when they jointly distinguish what is known from what is not, may never by explicitly considered.

Acknowledging Uncertainty About GH

How have physicians responded to the uncertainties presented by the prospect of administering biosynthetic GH to children who do not appear to be GH-deficient? According to the recent literature, the response seems to fit the pattern identified by Katz.

Discussions of biomedical research in the literature emphasize our limited understanding and the need for further (long-term) studies. For example,

> Only long-term follow-up will establish whether sustained treatment with growth hormone will increase the final height of these short normal children. Additional studies are needed to determine whether some of these children have subtle abnormalities of growth hormone or somatomedin structure, secretion, or action. These preliminary results in a small sample of very short children raise important ethical, clinical, and economic issues. Until we have more knowledge of its long-term effects and possible adverse actions in these children and more sharply defined criteria for the selection of patients and dosage requirements, the extrapolation of these findings to support indiscriminate treatment of short normal children with this potent hormone is premature and unwarrented.[5]

And in his advice to practicing pediatricians, Louis E. Underwood begins with the following:

1. GH is a potent agent with a variety of important metabolic effects. The long-term side effects of prolonged therapy in children who appear not to have GH-deficiency are unknown.
2. To date, there are no reliable methods for predicting which short normal children will respond to GH therapy.
3. The ultimate physical and psychological benefits of GH therapy in these children are not known.[7]

If representative, these statements suggest that when talking to each other, physicians are aware of many of the biomedical and psychological—if not sociopolitical—uncertainties surrounding GH treatment. (However, with one exception,[8] I have seen no mention of the dilemmas confronting clinical research. Although many articles call for further studies, there seems to be little recognition of the ethical barriers to conducting them on any but the very shortest children.)

But this generally admirable humility and candor does not seem to have been extended directly to patients. For example, after emphasizing uncertainty about the risks and benefits of GH treatment in points 1, 2, and 3, above, Underwood offers the practicing pediatrician the following advice in point 6:

> 6. Before prescribing GH for a purpose not currently approved by the FDA, the pediatrician should balance the perceived need for growth and the possible psychologic benefits if growth occurs against the cost of the hormone, the emotional burden on the family, and the risk of harm should the trial fail.

Although he may not have intended his words to be taken literally, Underwood implies that this precarious balancing falls entirely on the shoulders of the physician. There is no indication that the patient and/or his or her parents should be appraised of the risk, costs, and uncertainties of the matter and invited to share in the decision-making. The physician, like Dostoevsky's Grand Inquisitor, will assume the agonizing burden of choice and then deliver the final unambiguous result to the parents: "Yes, growth hormone is right for your child and I will prescribe it," or "No, your child doesn't need growth hormone (or won't benefit from it, or the risks and costs are too great, and so on)."

Physicians' failure to convey adequately the uncertainties and limitations of GH treatment is suggested by two recent articles discussing the psychological effects on children who are GH-deficient. The first, examining the psychological effects of relative "treatment failure," noted a discrepancy between children's perception of success of treatment and that of their physicians: "While most of the children in our study at best doubled their pretreatment growth rate, they were unable to view this change with the same enthusiasm as their physicians."[9] Although the reasons appear to be complex, one contributing factor may have been that the doctors' more "theoretical" doubts and uncertainties about the nature and extent of success were not adequately communicated to the patients and their families. The second study shows that depressive feelings in hypopituitary children receiving GH treatment "occur in part because inadequate physician–patient communication has resulted in unrealistic expectations of treatment, with subsequent disappointment when these expectations are not realized.''[11] Part of the difficulty, the authors suggest, ''may result from the infusions of hope and encouragement by the physicians in their previous encounters.'' Although they do not try to determine whether the breakdown in communication was attributable primarily to (a) an overly certain and optimistic picture painted by the physicians; (b) an inability to face the truth on the part of patients and their families; or (c) some combination of both, if Katz is correct (a), either by itself or in combination with (b), is likely to be an important factor.

Responding to Parental Concerns

Parents of otherwise normal children whose growth is constitutionally delayed or who, for genetic reasons, are likely to be short or shorter than they or their parents would like, are already making inquiries about GH treatment. Physicians should resist any temptation to make either of two quick responses. Given the risks, costs, and uncertainties noted above, it would be professionally irresponsible to begin treatment immediately. Nearly as problematic, to my mind, is belittling the treatment as merely cosmetic and the parents' interest as a sign of vanity. Parents' concerns about their children's height are often grounded in personal experiences or anecdotal or scientific evidence of various forms of discrimination. Dismissing their concerns too lightly may simply send them off to someone else who is only too willing to give them what they think they want.

The best response to parental concern is to take the time to discuss the matter thoroughly and to share, as fully and as clearly as one can, our knowledge of the risks, costs, and uncertainty about GH treatment. Once honestly appraised of this information most parents will, I suspect, be less enthused about treatment. Yet their anxiety and concern is likely to remain. Here the physician, nurse, or counselor may be able, through conversation, to be of more positive assistance.

The first thing to do is to acknowledge the legitimacy of the parents' concern. The mounting evidence for some degree of discrimination against short men and women must be taken quite seriously. It may then be helpful to compare this situation with the recent responses of blacks and women to various social prejudices. Rather than blacks trying to become white or women trying to become men, members of civil rights movement and the women's movement have worked to combat discrimination. Seen in this way, with best response to ''heightism'' in our society is not the unrestricted use of GH—which is in some respects a capitulation to this prejudice—but rather concerted efforts to reveal and combat it.

A parent may agree with this in general, but still be concerned about the welfare of his or her particular child. It may be many generations before discrimination against shorter men and women is fully revealed and significantly reduced. "Why," such a parent might ask, "should my child have to run this risk in the meantime?" In response, it might be pointed out that many short people are inwardly content and externally successful. What is their secret? According to John S. Gillis, the most important factor is the support they received from their parents:

> Most people are affected by how others view them, but children, in particular, experience enormous psychological pressures when they feel others see them as "too tall" or "too small." The support they receive, or do not receive, from their parents can make the difference between their beating, or being beaten by, these pressures.[3]

Gillis suggests a number of things parents can do to help their children overcome the possible disadvantages of being at either the lower or higher extremes of the height curve. Parents should first come to grips with their own anxieties about height: "Even parents of average height can have suprisingly strong emotional hangups about height." Then they should endeavor to foster in their children a strong sense of self-esteem. Emphasize the child's strengths rather than possible liabilities. The self-esteem that is built up through positive experiences in sports is accessible to smaller children through soccer, swimming, gymnastics, field hockey, and other competitive efforts that do not put a premium on height. The use of humor to ward off thoughtless or cruel comments from others can also be fostered. Furthermore, parents must remember to treat short normal children in accord with their actual age, not their height. Otherwise, they are likely to be overprotected and become self-conscious and timid. These and other sound tips offered by Gillis will both help the child and respond to the parents' need to do something for their child. Their interest in GH is partly motivated by a desire to take some sort of action, and Gillis's advice speaks to this desire.

Suppose, however, that a physician takes the time to share the costs, risks, and uncertainties of GH treatment and suggests alternative approaches for dealing with the problem, yet still the parents insist on the series of injections. Must the physician agree? Not at all. The physician's conception of sound medical practice and sense of moral and professional integrity are often sufficient to justify a refusal of such demands.[12] But it is important that such a refusal follows an honest disclosure and discussion of medical uncertainty and not be disguised as a false or overstated claim about the treatment's ill effects. It is one thing to refuse a request for treatment because it violates ones integrity or conception of sound medical practice; and quite another to do so on the basis of unwarranted claims to medical certainty.

Conclusion

The new possibilities created by the prospect of practically limitless supplies of GH are beset with uncertainties that should be frankly acknowledged by all parties—physicians, parents, children, and society. If advances in medical science have not eliminated uncertainty, they have often enabled physicians to distinguish more clearly between what we know and what we do not know: where we can be reasonably certain of a particular result and where the desired degree of certainty escapes us. Physicians are, as I have suggested, aware of many of the current clinical uncertainties of GH treatment. Now they need to be more forthright in sharing them with their patients and the public. And they need to become more sensitive to the psychological and sociopolitical uncertainties.

To acknowledge such uncertainties is not to abandon patients. On the contrary, it opens the way to the sort of interpersonal discussion and counseling that is, in many contexts, "the best medicine."

Acknowledgments

Portions of this essay are adapted from Martin Benjamin, James Muyskens, and Paul Saenger, "Short Children, Anxious Parents: Is Growth Hormone the Answer?" *Hastings Center Report,* 14 (April 1984), 5–9, and are reprinted with permission. I am also grateful to Ronna Benjamin and Joyce Bermel for a number of useful suggestions.

Notes and References

[1]J. Katz (1984) *The Silent World of Doctor and Patient.* The Free Press, New York.

[2]*New York Times* November 21, 1983.

[3]J. S. Gillis (1982) *Too Tall, Too Small.* Institute for Personality and Ability Testing, Inc., Illinois,

[4]For a more detailed account of the dynamics and treatment of short stature, see Paul Saenger (1982) Approach to the child with short stature. *Montefiore Medicine,* **7.**

[5] G. Van Vliet, D. M. Styne, S. I. Kaplan, and M. M. Grumbach (1983) Growth hormone treatment for short stature. *New Eng. J. Med.* **309,** 1016–22.

[6] J. M. Gertner, M. Genel, S. P. Gianfredi, R. L. Hintz, R. G. Rosenfeld, W. V. Tamborlane, and D. M. Wilson (1984) Prospective clinical trial of human growth hormone in short children without growth hormone deficiency, *J. of Pediatr.* **104,** 172–77.

[7] L. E. Underwood (1984) Growth hormone treatment for short children. *J. of Pediatr.* **104,** 237.

[8] Ad Hoc Committee on Growth Hormone Usage, The Lawson Wilkins Pediatric Endocrine Society, and Committee on Drugs, American Academy of Pediatrics (1983) Growth hormone in the treatment of children with short stature. *Pediatrics,* **72,** 893.

[9] D. Rotnem, D. J. Cohen, R. Hintz, and M. Genel (1979) Psychological sequelae of relative "treatment failure" for children receiving human growth hormone replacement. *J. Am. Acad. Child Psychiatry* **18,** 515–17.

[12] Department of Health and Human Services: Additional protections for children involved as subjects in research (1983) *Federal Register* **48,** 9814–20.

[11] R. S. Grew, B. Stabler, R. W. Williams, and L. E. Underwood (1983) Facilitating patient understanding in the treatment of growth delay. *Clin. Pediatr.* **22,** 685.

[12] M. Siegler (1983) Physician's refusal of patient demands, in *In Search of Equity,* (Bayer, R., Caplan, A. L. and Daniels, N., eds.), Plenum, New York.

Section IV

Concepts of Health and Disease

Introduction

The papers in this section grew out of a course team taught by Drs. Poirier and Bechtel (along with an historian, Dr. Daniel Jones) for students in the health sciences at the University of Illinois Medical Center. The point of the course was not to produce one unified perspective on the question of what constitutes health and disease, but to show a variety of insights the humanities can offer on this central question for the health sciences. A common thread between these papers is the issue of how a concept of health must be relativized to an individual's context.

Doctor Poirier's approach is to use literature to acquire an understanding of how people use the label "sick" when either they, or another person, reach a stage in the life cycle where we naturally expect the body to function less optimally. She considers three literary works that illustrate different nuances on one pattern for dealing with aging. In Laurence's novel, Hagar accepts the changes in her body as natural phenomena to which she must adapt, whereas her son and daughter-in-law construe as the problem these changes. Likewise, Barbara McDonald, in describing her own aging, shows an acceptance and appreciation of the changes occurring in her body, but faces a serious challenge in the responses others make to her aging. Finally, in Rathke's "Meditations of an Old Woman," the persona contemplates the changes in her body, and in that act, becomes more aware of herself as the agent somewhat removed from her body and so able to study it. Again, though, she must confront and overcome the perceptions others have of her. In Poirier's analysis, each of these women still thinks of herself as healthy; each has developed a new mode for adapting to her body. Their most serious challenges stem from younger people, who view age as disease and try to force these women into a sick role. This provides a far greater challenge to which the women must repond if they are not to succumb to illness.

Whereas Poirier focuses on the need for a dynamic concept of health that allows for the difference in the circumstances between individual's, Bechtel's endeavor is to find some objective foundation for thinking about health. He argues first that health, not disease, should be the basic concept, but he criticizes authors like Engelhardt and Margolis who have argued for a relativistic concept of health. Yet, he also departs from more traditional naturalistic concepts of health, such as those defended by Kass and Boorse who envision an ideal standard of health. With the relativistic critics of Kass and Boorse and with Poirier, he maintains that what is a healthy condition is environmentally relative. But he argues that a concept grounded in an evolutionary framework (generalized to incorporate not just biological, but also cultural evolution) can offer guidance as to what is a well-adapted state of an individual in particular environments. To further explicate his concept and to test its implications, he concludes by considering the consequences of such a naturalistic concept of health for the domain of genetic health.

In Defense of a Naturalistic Concept of Health

William Bechtel

Introduction

Many scholars have pointed out that defining the concepts "health" and "disease," which may seem to be nothing more than an abstract, intellectual enterprise, has important practical consequences. These consequences determine how the health care profession is practiced and the manner in which it affects the lives of its clients. In the first place, including someone under the definition of the terms "healthy" or "diseased" has consequences for the way in which we treat that person. Although Parson's concept of the "sick role" may require modification in light of recent developments in our culture (for example, there is a growing sentiment that in many cases an individual is responsible for being sick), we still exonerate people from a number of social obligations when they suffer from disease, including employment obligations and responsibilities for behavior.[1] As well, we expect those who are exonerated to seek professional help(and increasingly, we expect society to provide them such help). Hence, how we define the concepts of health and disease determines to whom we grant such exonerations.

Second, how we define the concepts determines what problems health-care workers will address. Engelhardt enumerates a number of conditions that may or may not fall within the domain of health care, depending on the definition of concepts of "health" and "disease": alcoholism, homosexuality, menopause, and aging.[2],[3] Szasz invokes a particularly narrow conception of disease to eliminate from the domain of disease many conditions now classified as mental illnesses.[4] At the other extreme, there is the account of a physician in the South who wrote a prescription for a tractor, viewing it as essential for the nutritional health of the group under treatment. Gossens takes this view to an even further extreme, arguing that medicine ought to be concerned with well-being and not limited to health and disease.[5]

Finally, the concepts of ''health'' and ''disease'' that are adopted also determine how health-care workers define their mission. Dubos describes the continuing conflict between the Hygeian and Asclepian models in health care.[6],[7] The Asclepian model defines health as the absence of disease and construes the mission of health care as combating disease. The Hygeian model of health, in contrast, views health as compatible with the survivial of disease entities. It presents health as involving a kind of balance in one's relationship to one's environment, and construes the mission of health care as maintaining this balanced relationship. As later discussion will show, these interpretations of the mission of health care are quite different.

Given that ''health'' and ''disease'' play such important normative functions in our lives, we need to address the issue of whether we can justify invoking one concept of health and disease rather than another. This issue quickly gives rise to others. Like many other normative terms, the concepts ''health'' and ''disease'' also have a descriptive function. One can characterize the state of a person by indicating that he or she is suffering from a certain disease or is healthy. This poses the question of whether the descriptive function or the normative function of these terms is most basic and determines the other. Scholars have differed on this issue. Kass, for example, thinks that there are objective criteria for the descriptive use of the term ''health,'' and that the normative use of the term depends on this descriptive content of the terms.[8] Engelhardt, on the other hand, contends that the normative function of the terms is primary and determines the descriptive extension of the terms.[9]

Coupled with this issue is the question of which of the concepts (''health'' or ''disease'') ought to take primacy. Most scholars take the terms ''health'' and ''disease'' to be correlative, so that health is considered to be the absence of disease, whereas disease is construed as a detraction from health.[10] However, showing that the terms are correlative does not do away with the need to consider which is primary. Traditionally, two conflicting traditions (closely aligned with the Hygeian and Asclepian traditions noted above) have pursued each of these alternatives. One of these holds that it is possible to define the class of disease entities independently, and the other maintains that we must define health first, since we can only identify diseases as deviations from such a reference point.

In this paper my ultimate objective is to explore whether, within the framework of contemporary biology, one can make coherent a position on the first question (the question of whether the normative or descriptive function of the concepts "health" and "disease" is primary), which gives primacy to the descriptive function of the concepts of "health" and "disease." To pursue this issue, however, I will need to enter into the debate over the second issue (the issue of whether "health" or "disease" is the primary concept).. On this issue, I shall argue that "health" is the primary concept and that "disease" is derivative. Although numerous others have followed this route, both my reasons for taking it and the concept of health that I shall articulate differ from those given by others.

A coherent position is not necessarily the correct position. I will not try to develop a definitive argument that the position I explore here is correct. The points I will discuss suggest that this position should be given serious consideration. One way to evaluate it is by exploring its consequences for the practice of health care. To the degree that these consequences seem plausible, the position is further supported; to the degree that they violate our intuitions without providing good grounds for modifying those intuitions, the position is weakened. (The evaluation procedure I am proposing here is reminiscent of that proposed by Rawls for basic moral principles).[11] I will conclude this paper by exploring briefly the implications of this position for the domain of genetic health. I have chosen this case because genetic health appears to be a difficult case for most concepts of health and yet one that demands our attention. My comments there are far from definitive; they are intended to open the inquiry, not close it.

One final comment should be made about the kind of definition of "health" and "disease" to be advanced here. It will be a definition that places these concepts in a theoretical matrix. To classify someone under one of these terms will require invoking this theoretical matrix, not just examining the individual to be classified. Some parts of this theoretical network may not be precisely defined at present, which will render application of the terms "health" and "disease" somewhat uncertain at present. I do not consider this to be a negative consequencee of definition, as long as the definition makes clear what kinds of theoretical advances are needed to resolve these uncertainties. Another consequence of embedding these terms within a theory is that their application may change either as the theory is revised or as the theory is applied in different contexts. This, however, as we will see later, is a virtue, not a defect, of this approach.

The Priority of the Concept of "Health"

As we saw above, even accepting that the terms health and disease are correlative, there remains a serious issue as to which of these concepts ought to be construed as basic. Historians of medicine have distinguished two quite different traditions—the ontological tradition and the physiological tradition. (The distinction goes back to Broussais).[12] The ontological tradition treats diseases as clearly identifiable entities, and construes the task of health care to be the curing of these diseases. The physiological tradition, in contrast, gives primacy to health, understood in terms of the proper functioning of the individual. It then construes diseases as deviations from this condition. Representatives of the physiological tradition have typically been nominalists about disease. That is, they do not believe that there is a common essence to various conditions for which a particular disease name is used, and they view the names themselves as simply providing useful shorthands for describing a range of conditions.

Numerous scholars have traced the development of these two conceptions through the history of medicine.[13],[14] This discussion will focus more on the conceptual problems each approach faces. Insofar as we shall discuss historical uses of these two frameworks, it will be to advance this conceptual discussion, not to provide a rigorous historical analysis. In this section I will argue for a physiological conception by pointing out difficulties with the ontological conception. This, of course, does not show that the physiological conception fares any better. The following section, though, will indicate some virtues of the physiological conception by showing how, working within it, one can develop a coherent naturalistic conception of health (i.e., one that treats the term "health" as having a determinate descriptive content and not as serving solely as a evaluative term.)

Engelhardt distinguished two orientations that the ontological approach can take.[14] The first involves simply classifying disease conditions within a loogical type, whereas the second assigns the etiology of these conditions to a disease entity. I will discuss these two orientations in turn and show the difficulties each encounters.

The first orientation is reflected by such physicians as Sydenham and Pinel who tried to develop extensive toxonomies of diseases, called nosologies.[15,16] In Sydenham's work, it is clear that these taxonomies are based on the commonality of symptoms: "Nature, in the production of disease, is uniform and consistent; so much so, that for the same disease in different persons the symptoms are for the most part the same. . . ."[17] The objective for Sydenham was to identify modes of treatment that would be efficacious for instances of a disease category, such as the use of chincona bark for treatment of malaria.

One of the difficulties facing the nosological tradition is that disease conditions in different individuals are not identical. To handle this, those offering nosologies tried to distinguish essential symptoms from accidental or secondary characteristics. Such distinctions, however, were controversial. Critics objected that the diseases distinguished by the nosologists were not real, but artificial, resulting from forcing individual disease conditions into specially concocted types in order to place them within a classification scheme. Thus Broussais criticized the nosologists: "One has filled the nosological framework with groups of most arbitrary formed symptoms . . . which do not represent the affections of different organs, that is, the real diseases. The groups of symptoms are derived from entities or abstract beings, which are most completely artificial *ontoi;* these entities are false . . ."[12]

For nosologists to answer this objection, it was necessary to show that the variety of symptoms appearing in different cases of a disease type could all be explained in terms of a common basic condition. Pathological anatomy significantly assisted nosologists in this cause. Anatomists, beginning with Benivieni, began to identify specific lesions associated with various collections of symptoms.[18] Morgagni provided a systematization of the results of numerous *post mortem* analyses, thus providing the basis for identifying lesions underlying the diseases included in various nosologies.[19] Bichat brought the discipline of pathological anatomy to a sufficiently high state of development that it could associate a wide range of diseases with specific underlying lesions.[20] This search for a common pattern was helped by techniques that were developed to identify lesions in living patients[21] and by the advent of hospitals that made available large populations of patients in which one could look for and identify common disease patterns.

Classification schemes can have great utility. By placing an object or condition within a type, we can use information about other instances of the type to predict and perhaps control what will happen in this instance. The same holds true for disease conditions. Identifying a particular patient as suffering a disease of an established type enables one to use what is known about that type of disease to make a prognosis. Knowing how successful particular treatments have been with previous instances of the condition, one is able to select from various treatments. Although I will later raise some cautions about even this use of the ontological concept of disease, and also show that it cannot stand independently of a concept of health, for the most part this rendering of the ontological perspective has performed a salutory function in organizing the knowledge health-care workers require in order to diagnose and treat patients (for a more negative appraisal, *see* Engel).[22]

Most of the objections to the ontological concept of disease have been directed at the second rendering of the ontological concept which traces the etiology of disease states to disease entities. This version of the ontological conception has roots further back than the nosologies I have been discussing. It can be traced back to folk medicine and the idea that disease is the invasion of the person by a spirit. This perspective was provided a naturalistic interpretation when Fracastoro introduced the concept of contagion.[23] The contagion notion explained epidemics such as the plague as resulting from imperceptible particles that were themselve the disease entities invading the body. Paracelsus von Helmont, and other founders of modern medicine, adopted the idea that each disease is caused by a specific disease entity; Koch's discovery of the tubercle bacillus made it the foundation of the popular germ theory.[24] According to this theory, identifying the responsible microorganism is a prerequisite for determining that a constellation of symptoms is a disease.

The germ theory also inspired a concept of how to treat diseases: Eradicate the invading microorganism. This resulted both in the quest for antibacterial agents and the advent of immunology. The successes of the health-care professions in treating diseases through such procedures has given this version of the ontological conception such credibility in the 20th century that it has become the foundation of our folk beliefs and, as such, has seemed virtually undeniable.

The domination of this version of the ontological conception of disease, that defines disease conditions in terms of an invading microorganism can be seen in the early work on avitaminosis diseases. The co-occurrence of symptoms associated with beriberi suggested that it was a single disease, leading germ researchers to seek a bacterial agent to explain the condition. When Eijkman discovered that beriberi could be avoided or cured by substituting unpolished rice for polished rice, he proposed that the kernel of rice contained a toxin, whereas the husk, which was removed in polishing rice, contained an antitoxin.[25] Only in 1905 did Eijkman accept Grijns' proposal of 1901 that beriberi was caused by the lack of a vital nutrient in the rice kernel.[26] Researchers resisted the idea that the deficiency of a nutrient, especially one that is present in only minute amounts in a normal diet, could cause a disease condition.The idea was not accepted until the works of Hopkins and Funk, who showed that several diseases fell into this category, and introduced the concept of a vitamin for necessary nutrients in minute quantities whose absence could produced diseases.[27-29] (I have discussed the process through which the vitamin concept was established in more detail elsewhere).[30]

The difference between the vitamin theory and the germ theory is that the vitamin theory holds that avitaminosis diseases result from the shortage of the vitamin, rather than from the presence of a germ. Although the discovery of vitamins undercuts the assumption of the germ theory that an external agent was required for each disease, it provided additional support for the more general ontological conception of disease; the idea that a specific entity is responsible for each disease. More recent challenges, however, have attacked this one entity–one disease assumption by trying to show that disease entities do not work alone in producing disease conditions. This is demonstrated by the fact that infectious diseases do not affect everyone alike. Some people become violently ill when a disease entity attacks their body, others become mildly ill, and still others are not overtly affected at all. Dubos uses the setting in which Koch presented his claim to have traced tuberculosis to the tubercle bacillus to raise this objection: "It can be stated with great assurance that most of the persons present in the very room where he read his epoch-making paper in 1882 had been at some time infected with tubercle bacilli and probably still carried virulent infection in their bodies. At that time, in Europe, practically all city dwellers were infected, even though only a relatively small percentage of them developed tuberculosis or suffered in any way from their infection."[6] Dubos presents evidence that Koch himself was affected. (*See* Engel for a discussion of how different individuals vary in their responses to the tubercle bacillus.)[22]

Critics of the ontological conception, like Dubos, view cases such as this as undermining the link between diseases and disease entities, and as grounds for an interactive concept of health and disease. The fact that individuals exhibit different responses has not been ignored by defenders of the ontological conception, however. In fact, they explain this phenomenon in the same manner as their critics; namely, by appealing to differences in internal constitution that affect individuals' susceptibility to germ attacks. Although conceding that differences in internal constitution affect what symptoms the persons will manifest, they justify defining the disease in terms of the invading microorganism (or other entity) by arguing that it is distinguishable as the proximate cause of the disease condition.

By treating the invading microorganism as the proximate cause of the disease condition and identifying the disease with it, the ontologist gives it primary focus and reduces the other factors to the less significant status of background conditions. The power of this assumption is apparent when medical researchers confront disease conditions in which a specific causal factor has not yet been identified. Dubos shows how contemporary researchers, in dealing with cancer, arteriosclerosis, and mental disorders, are following the prescription of the ontological tradition and seeking *the* proximate cause for each condition.[6]

One consequence of the one entity–one disease assumption is that whenever one finds different causes for what appears to be the same disease, one must distinguish between diseases. Historically, this was done in the case of tuberculosis.[3] However, in some cases another tack has been taken. For example, as Dubos recounts, the early experimental research on diabetes traced it to a deficiency in the pancreas that reduced insulin production. However, the symptomology of diabetes can be equally produced by lesions quite remote from the pancreas, and patients suffering from these other lesions are still considered to have diabetes. Perhaps one reason these conditions continue to be classified together is that they can all be treated by insulin therapy. What is significant about this case is that health-care researchers accepted a one-to-many mapping of a disease condition into underlying etiologies. The tendency to refer to disease syndromes (e.g., General Adaptation Syndrome), in which one often finds a cluster of disease conditions that result from a variety of underlying causes, also shows an acceptance of one-to-many mappings.

The critical assumption of the version of the ontological concept of disease I have been discussing (namely the assumption that there is a one-to-one mapping of disease conditions onto underlying etiologies) deserves further discussion, because it is not an isolated assumption, but is a feature of a powerful explanatory strategy. This strategy directs one who is interested in explaining a particular phenomenon to isolate and so localize discrete factors that may be causally responsible for it.[31] Scientists have developed powerful tools to carry out such localizations and they have played a valuable role in the advance of science. However, they have also led scientists into a kind of error that I have identified elsewhere as a functional localization error. The error consists in inferring that a localized factor that may be contributory to the phenomenon in question is *the* cause of it. By making this inference, one ignores the interaction of factors involved in producing the phenomenon (*see* Engelhardt and Erde for a discussion of how

Virchow struggled with this problem as it was posed by the theory of contagion).[35]

Identifying diseases with underlying etiologies is an instance of a functional localization assumption that can have critical consequences, for example, when one is trying to design a course of therapy. Spaeth and Barber discuss a case in which they claim such consequences have accrued.[36] In 1962, homocystinuria was identified as a disease with a variety of symptoms (dislocated lenses, glaucoma, mental retardation), but with only one underlying lesion—the deficiency of pyridoxine resulting from a defective gene. As a result of accepting this localization assumption, Spaeth and Barber claim researchers ignored and delayed looking for other factors that now appear to constitute alternative etiologies for homocystinuria. In this case, the localization assumption affected the research endeavor, preventing scientists from understanding the different etiologies that might be involved in cases that appear quite similar.

Localization assumptions can have similar consequences for those trying to choose courses of therapy. If one is cognizant that there are or might be a nexus of causal factors responsible for the disease, one is in position to identify alternative modes of treatment. Thus, to eradicate the disease efficiently, one could turn to that part of the nexus that is most readily under control. This may not be the microorganism or other disease entity. But when the disease is identified with the disease entity, one may be blinded to these other courses of therapy. Dubos argues that, in fact, diseases like the plague were controlled not by directly attacking the responsible microorganisms, but by a variety of public health measures that served to limit the impact of the responsible microorganisms on human health.

The fact that the one disease–one entity model often results in a functional localization error constitutes a serious objection to the version of the ontological concept of disease that links diseases with underlying disease entities. The hazards of the localization problem are avoided with the interpretation of the ontological position discussed earlier which only sought to identify the disease condition and not to identify it with a disease entity. That version can treat the microorganisms that are the focus in the second concept as merely one part of the causal nexus that results in the disease condition and so avoid focusing on it alone.

There remains, however, an objection to either version of the ontological concept that, for the purpose of this paper, seems even more significant than those just discussed. Even assuming that health-care professionals can differentiate disease conditions into types, or kinds, in a consistent way, the principal issue being investigated here is what makes these groups or kinds of diseases. Another way to state the issue is: Is disease itself a kind? In most taxonomic hierarchies, kinds at one level constitute members of kinds at a higher level (at what Rosch calls the superordinate level).[37] Even if there is no essence (a characteristic that is necessary and sufficient for being a member of the kind) to such higher level taxa, a taxonomic scheme requires some kind of unity among the various conditions satisfying the superordinate term, perhaps the kind of coherence found in fuzzy sets. Thus, we ought to be able to specify criteria that are relevant to determining whether something is a disease. Moreover, if disease is the primitive notion, these criteria cannot be negative, as would be the criterion of being disruptive to health.

It seems clear, once the demand is put in this way, that the ontological conception cannot claim primacy. Szasz's strategy in 1961 of defining diseases in terms of lesions seems the most plausible. However, this will not do, because one must, in turn, explicate the notion of a lesion. One usually construes a lesion as a localized damaged state. However one requires knowledge of the healthy state in order to assess whether there is damage. To avoid such an appeal to health in defining disease, one must identify some intrinsic characteristics of disease states. Moreover, these characteristics should be projectable to new instances. One cannot prove the negative claim that no such set of characteristics exists that defines the concept of "disease," but it seems highly implausible that there is such a set, given the wide variation observed in disease conditions.[38] Rather, when one surveys the range of conditions that get classified as diseases, it seems natural to think of these conditions as deviations from states of health.

Given that the onological conception cannot stand alone, we need now to consider whether it is possible to define "health" first and then define "disease" in terms of it. It should be noted that rejecting "disease" as the primitive concept does not undermine the endeavor of pathologists to develop nosologies or even to identify diseases with disease entities (although that might not be advisable, given my earlier discussion). What it does claim is that "disease" must be understood as the complement of "health," and that priority must be given to defining health. Disease entities can then be defined as particular deviations from health if sufficient regularity is found. It is in the physiological tradition that the concept of "health" is primary and "disease" is defined in terms of it. It proposes to define "health" in terms of theories about the normal or proper physiological processes in the organism. (The term "physiological" is construed broadly to include any kind of processes in the organism, thus including biochemical processes and psychological processes.) Although this endeavor seems well motivated, a major difficulty surfaces immediately: What is to be the standard of normalcy or properness in terms of which the healthy condition is defined? In the next section I will explore the approaches one might take in establishing these criteria upon which the physiological concept of health depends.

Establishing a Naturalistic Interpretation of the Physiological Concept of Health

The challenge faced by the physiological tradition is to specify what state of an organism counts as healthy. The strategy typically followed is to view health as an enabling condition; a condition that allows one to participate in other activities. This represents one half of Galen's account, for Galen identified health as a condition "in which we neither suffer pain nor are hindered in the functions of daily life."[39],[40] The first aspect of his account, not suffering pain, has not assumed as much prominence, probably because pain is itself a hindrance to performing the functions of life. Connecting health with being able to perform the functions of life, however, does not alone solve the challenge, for it is necessary now to specify the functions of life.

The attempt to explicate the notion of function for human beings has lead to a central disagreement between two groups, both of whom defend a physiological concept of health. The dispute concerns whether the function of an organism, particularly of a human being, is naturally determined or depends on the values of a society or particular individual. This issue is part of a broader dispute, because "function" is a teleological notion and there is a long-enduring dispute over whether there is a place for teleology in modern biology.[41] One side in the dispute accepts a teleological perspective as central to biology and believes that normal functioning (as it is relevant to evaluations of health and disease) can be determined in a naturalistic way. The other side rejects both a naturalistic teleology and the idea that a naturalistic determination of proper functioning is possible. It holds, instead, that proper functioning can only be socially (or, in the extreme version of the position, individually) defined.

My strategy in this section will be to review both the traditional defense of and opposition to a naturalistic teleology. Although my sympathies lie with the first group, I find their defense inadequate. Therefore, I will conclude this section by developing an alternative defense of a teleological perspective within modern biology (one that does not violate the canons of mechanism) and exploring how it offers a framework for a naturalistic concept of health.

The Traditional Naturalistic Interpretation

Kass and Boorse have recently offered defenses of the traditional naturalistic version of the physiological conception of health.[8,42] Kass appeals to the notions of "wholeness" and "well-working" in explicating his naturalistic conception: "Health is 'the well-working of the organism as a whole.' " Part of Kass' attempt to justify his use of the concept "wholeness" consists in pointing out that living organisms constitute integrated wholes that exhibit homeostasis (points to which I will return in the last part of this section). The crux of his case, however, is his conception of a well-working organism. To judge this, he appeals to our intuitions as to when a living system (e.g., a squirrel) is working well for the kind of organism that it is: "The healthy squirrel is a bushy-tailed fellow who looks and acts like a squirrel; who leaps through the trees with great daring; who gathers, buries, and covers but later uncovers and recovers his acorns. . . . "[8] Kass, however, does not provide any additional criteria other than our intuitions for measuring whether an entity is whole or well-working. From the tone of his presentation, one gathers that Kass assumes a form of Aristotelian essentialism that supports such judgments. Such a perspective might have seemed plausible within a nonevolutionary world view, but it is particularly suspect within an evolutionary framework that rejects essentialism and sees species as continually changing under such forces as natural selection.[43]

Boorse provides a good deal more guidance for determining when something is functioning properly. He proposes grounding the determination on medical theory: " . . . behind this conceptual framework of medical practice stands an autonomous framework of medical theory, a body of doctrine that describes the functioning of a healthy body, classifies various deviations from such functioning as diseases, predicts their behavior under various forms of treatment, etc. This theoretical corpus looks in every way continuous with theory in biology and the other natural sciences, and I believe it to be value-free."[42] I concur with the predilection to think it is possible to develop a theory that differentiates healthy from diseased functioning on a naturalistic basis; the challenge is to articulate it.

Ultimately, Boorse appeals to a standard of normality to distinguish healthy from diseased functioning. First, however, he gives the impression of trying to explain what normal is. He cites a statement from King that appears to define normal in terms of the design of the organism: "The normal . . . is objectively, and properly, to be defined as that which functions in accordance with its design."[44] This, however, will not work, because it puts Boorse in the same predicament as Kass, needing criteria for determining what something's design is. As I noted above, within an evolutionary framework, one cannot equate the design of something with its internal essence, because the organism is changing under selection pressures. But Boorse thinks he has a way out. He says "a function . . . is nothing but a standard causal contribution to a goal actually pursued by the organism." These goals, in turn, can be ascertained empirically, "without considering the value of pursuing them."[42] This is accomplished by studying the normal behavior of the organism, with normal taken in the statistical sense. Boorse has rather clearly moved in a circle now, for normal is being used to define functioning acording to design, which was to explicate the notion of normal.

Moreover, as Boorse is aware, there is a difficulty in defining health in terms of statistical normality. Such a move makes a condition such as caries, which is widely distributed in our species, a healthy condition, when, pretheoretically, it ought to count as a disease.[45] He thinks he can avoid such anomalies by adding an additional criterion that makes a condition count as diseased if brought about by external causes. Thus, he offers a general definition of disease: "Deficiencies in the functional efficiency of the body are diseases when they are unnatural, and they may be unnatural either by being atypical or by being attributable mainly to the action of a hostile environment."[42] (If this modification is to solve the above problem, Boorse needs to amend this to hold that a disease is an abnormality that is atypical *and* caused by the action of a hostile environment.) To make this definition work, however, we need to know what actions are hostile, something we can only judge once we know the state they are disruptive of and hence hostile to.

To defend his position, Boorse offers an analogy that, contrary to his intent, seems to highlight the fundamental difficulty with his view. He appeals to artifacts like a Volkswagen in which he claims "properly functioning" is a purely descriptive term. In the case of artifacts, however, we have independent access to their design in terms of which we can evaluate a particular instance. We need to know how to acquire such knowledge in the case of natural objects. It is not sufficient to assert, as Boorse does, that when a car is in production, it is an empirical matter what the design is. It is still only because one can appeal to the design specifications or the design objectives that one is able to make such judgments. Without knowing either the design specifications or the design objective, and having only a collection of more or less properly running Volkswagens to inspect, we would have no basis for determining the design to which they were supposed to conform. One needs to be particularly careful in appealing to artifacts when trying to judge the status of something's design, because with artifacts we have a privileged access to design information. The challenge in the case of nonartifacts is to ascertain what the design is, for it is in terms of it that we judge whether the environment is hostile.

Although I have raised a number of objections to the details of Boorse's analysis, I remain sympathetic to his initial intuition that medical or physiological theory might provide an adequate grounding for an analysis of health. The idea is that by examining how the organism operates we can discover standards for its proper operation. The problem, however, is that physiological theory does not seem to support such determinations. Galenic medical theory, insofar as it includes teleological elements, seems to come closest to providing such support, for it postulates a state of health as arising when the four humors are in proper equilibrium. However, this teleological perspective is imported into the physiological theory proper, for the equilibrium state is ultimately determined by when the body is prepared to exercise its normal functions. For Galen, these normal functions include being able "to take part in government, to bathe, drink, and eat, and do the other things we want."[39] For Galen, it was not the case that knowledge of the healthy equilibrium state determined the natural functions of a human being. Rather, the determination of the healthy equilibrium state itself came from a perspective of when the organism seemed to be thriving (which, in Galen's case, was provided by the cultural ideals of the Greeks).

Modern physiology, moreover, seems to offer even less hope than Galenic physiology for grounding an account of proper functioning. Modern physiology seems to have conceived of the body as a mechanism and, as such, something that can be manipulated so as to serve whatever functions (within limits) to which one wants to put it. Later we will see that this mechanistic concept of physiology is incomplete, and that when it is integrated into a more comprehensive biological perspective, there is a basis for a naturalistic determination of proper functioning.[46] One of the major things that is required is an integration of physiology with evolutionary theory, something not proposed by either Kass or Boorse. This integration, however, at first seems to work against a naturalistic interpretation. Many of the critics of the naturalistic perspective have made use of an evolutionary framework in their critique. Therefore, I will delay developing my positive proposal until after I have examined this critique.

Opposition to a Naturalistic Interpretation

The difficulties I have been noting with a naturalistic interpretation of the physiological concept of health stem from the need it imposes to specify proper human functioning in a principled (not merely intuitive) way. The evolutionary framework seems to undermine this, for in an evolutionary scheme, proper functioning involves being adapted to one's environment. But such adaptedness in evolutionary terms is environment-relative. Because different environments make different demands on organisms, a condition that may be adapted to one environment and contribute to the organism's functioning in it may be disadvantageous and render the organism disfunctional in another. Hence, Engelhardt claims: "An evolutionary appreciation of human functions makes functions dependent upon particular environments. Thus, since there are no standard environments, there can be no standard definition of human functions."[2] Dubos gives some examples that graphically illustrate this: "A Wall Street executive, a lumberjack in the Canadian Rockies, a newspaper boy at a crowded street corner, a steeplechase jockey, a cloistered monk, and the pilot of a supersonic combat plane have various physical and mental needs. The imperfections and limitations of the flesh and of the mind do not have equal importance for them. Their goals determine the kind of vigour and resistance required for success in their own lives."[7]

Sickle cell trait is perhaps the best-known example of a condition that in one environment increases an organism's fitness, but in another is either negative or neutral. Perhaps more common is the reverse situation in which a generally recognized disease may or may not be expressed, depending on environmental circumstances. Thus, Dubos claims: "Overwhelming evidence indicates [that] many forms of disease have emerged or have been disseminated in the modern world because our ways of life have created new and complex constellations of circumstances favorable for their spread,"[7] He offers as an example porphyria, a genetic condition that has historically produced only mild symptoms, but now results in violent reactions and death when those afflicted with it encounter sulfa drugs and barbiturates. There are, as well, numerous examples, especially from the expeditions that explored the world in recent centuries, in which one population has developed an immunity to a disease entity that ravages another population without the immunity.

The response that the defender of naturalism, Kass, has offered to this environmental relativity of health is inadequate. Kass maintains that a person is diseased, even if the disease does not manifest itself in a detrimental way in the current environment. Such examples of differential affects of diseases, he claims, "do not prove the relativity of health and unhealth. They show, rather, the relativity of the *importance* of health and unhealth. The person without hay fever, enzyme deficiency, myopia, paraplegia, and ingrown toenails, is, other things being equal, *healthier* than those *with* those conditions."[8] What Kass is trying to do is evaluate proper functioning not in one environment, but in the full range of possible human environments. However, it is doubtful whether sense can really be made of this requirement. The same condition that equips one to deal well with one environment may make one function poorly in another environment. Sickle cell trait (as opposed to the disease), for example, equipped individuals to resist malaria in Africa, and so improved their functioning. In low-oxygen environments, however, it can cause blackouts, a fact that has been used to keep those with the trait out of professions such as that of pilot.

Those who appeal to the environmental relativity of assessments of function to reject a naturalistic concept of natural functioning often claim that assessments of proper functioning depend on a particular individual's or society's values.[47] In support of this view, it is often pointed out that societies have, historically, differed in their assessment of what characteristics make one healthy or ill. Engelhardt and Engelhardt and Erde have enumerated a number of conditions that have been treated as cases of mental illness in different cultures.[2,14,35] For example, the desire of a slave to escape was considered an instance of mental illness (drapetomania) during the period of slavery in the United States, but in contemporary societies is considered a normal, healthy desire. These critics of the naturalistic perspective maintain that there is no way to get beyond these differences in values to one absolute standard of proper functioning.

However, examples similar to those offered to support social relativism in defining health can also be invoked to argue for the need for a naturalistic concept. The reason is that such relativism seems to provide an opening for abuses, such as those of the Holocaust or of Soviet psychiatry. If ideology can determine what constitutes health then, unless there are grounds for rejecting certain ideologies, there are no grounds for rejecting particular treatments of human beings in the name of promoting health if they conform to a socially accepted ideology.

The challenge is to develop a framework that allows for the environmental relativity of health that the critics of naturalism have appealed to without providing an opening for the abuses permitted by such appeal to values. Below I will develop such a framework. Before turning to that, however, there is another objection offered by

Margolis that should be considered.[48] He argues that, on logical principles, judgments of health and disease must be considered value judgments, for they involve ranking or grading a particular case against established norms. He presents the apparently nonnormative features of biomedical science as an abstraction from the normative framework: "The allegedly scientific and value-neutral status of medical pathology addressed to cells, organs, and biochemical processes, and the like, must be an abstraction (entirely defensible as such) from the value-freighted investigations of the world of disease and illness common to pathologist and clinician."[48]

In itself, this objection is not very telling, since it consists merely in pointing out that evaluations are always made against some standard. This is true of any predications, for any predication involves a judgment that a particular instance meets the criteria for satisfying the predicate. If one accepted Margolis' argument, then any predication would constitute a value judgment. The real issue, and the point underlying Margolis' challenge, concerns the status of the criteria by which normal and diseased cases are judged. By raising this issue, Margolis poses the challenge to the defender of a naturalistic conception; he does not settle the issue against the naturalistic conception.[49]

In addition to critiquing the naturalistic conception, Margolis tries to develop a concept of health that is not as relativistic as Engelhardt's. He notes first that, even if health is an intrinsically normative concept, it can take on a determinate descriptive function as it is employed. He suggests that the appropriate descriptive content for the concept "health" is determined by the shared prudential values of a species. These values constitute "enabling interests, that is, the general (determinable) condition on which any ethical, political, economic programs viable for a complex society must depend."[48] Thus, for Margolis, the criteria of health and disease are determined by reasoning back from actual normative preferences (ideologies) to those basic elements necessary for pursuing any ideology: " . . . medicine is ideology restricted by our sense of the minimal requirements of the functional integrity of the body and mind (health) enabling (prudentially) the characteristic activities and interests of the race to be pursued. And disease is whatever is judged to disorder or to cause to disorder, in the relevant way, the minimal integrity of body and mind relative to prudential functions."[48]

If there were such a set of prudential values, Margolis' position might provide a way out of some of the abuses relativism seems to engender. There are, however, some shortcomings with his position as its stands. He has not demonstrated that there is a common set of prudential values compatible with all ideologies, a formidable task to say the least. Further, we need an argument for limiting the scope of health to the provision of these prudential values. Rather than trying to overcome these difficulties, I suggest that greater progress will be made by developing a naturalistic framework that overcomes the objections raised to those offered by Boorse and Kass. Hence, I will now turn to sketching a more adequate naturalistic conception.

A Naturalistic Teleological Orientation

The character of the alternative naturalistic concept of health that I will be developing in this section can be put into perspective by a brief discussion of the history of physiology. Earlier we saw that the mechanistic physiology that emerged in the modern world seemed to exclude teleology and so offered no basis for determining what condition constituted the proper functioning of the organism. However, a mechanistic approach was not universally accepted by the early developers of modern physiology. Many continued to advocate some form of vitalism, arguing that there is a special character to the functions of life. Today these vitalists are often viewed as obscurantists who tried to block the path of scientific progress. This judgment is mistaken. The vitalists recognized an important feature of living systems—that they are organized systems and not just random assortments of physical parts. Often their appeals to vital powers were simply appeals to recognize the critical role played by this organization (*see* Lenoir for a discussion of the persepctive taken toward organization by several of the major figures in the development of physiology and how, over time, these investigators adopted naturalistic approaches to discussing organization).[50] Mechanists tended to neglect this organization. Ultimately, however, they have had to take it into account, and in so doing have granted a key contention of the vitalists.

A critical contribution made by Bernard was to show how one could take seriously the importance of organization within a completely naturalistic, mechanistic physiology.[51] Bernard introduced the notion of two environments: organs or cells within a living system live directly in what he termed an "internal environment," consisting of other components of the living organism, and only indirectly and by virtue of being component parts of an organism, in the "external environment," consisting of the world outside the ogranism.[52] In his exploration of the inner environment, Bernard showed that organisms were organized systems that were designed to maintain themselves in a basically constant state.[53] Cannon supplied the name "homeostasis" to this phenomenon.[54]

The idea that living organisms incorporate a complex organization that makes them homeostatic systems provides an important element needed in a satisfactory physiological concept of health. In terms of it, one can define a healthy system as one that is at or near its designed equilibrium state. Significant deviations, especially those in which some external agency is required to restore the system to the equilibrium state, are disease states. One virtue of this framework is that it permits one to explain the basis for the second objection to the ontological conception—that not all organisms in which a disease entity is present appear to be afflicted with the disease. This can be explained in terms of differences in the internal environment of different individuals. This framework can also explain another phenomena that presented a difficulty for the ontological concept—that different factors can give rise to the same disease conditions. This is explained by the fact that all of them result in the same kind of disruption of the internal environment.[55]

This scheme, however, is incomplete in a critical way, for it does not show why any premium should be placed on maintaining homeostasis. A useful perspective on this problem that reveals the nature of a potential solution can be found by further examining the history of physiology. One point on which both the vitalists and Bernard insisted, and on which they erred, concerns the origin of organization. They maintained that the mode of organization found in any living system had to be accepted as a starting point for inquiry and could not itself be explained. However, during the same period Darwin was formulating another mechanistic scheme, that of natural selection, that could explain the origin of organized living systems. Unlike other mechanistic schemes that exhibited a reductionist aura, Darwinian natural selection does not try to reduce the significance of organization. Rather, it is uniquely capable of revealing the importance of organized systems, for it shows that such systems have been selected because of their ability to meet environmental demands. Homeostatic systems are those that can survive and replicate in the face of fluctuations in their environments.

The importance of the evolutionary framework is that it provides a means for introducing a teleological perspective into biology, a perspective I will employ to complete the concept of health that I began to develop in discussing homeostasis. The above statement may seem paradoxical, because Darwin is often viewed as driving teleology out of biology. However, Wimsaatt and Wright have developed a quite different interpretation.[56],[57] I will briefly sketch Wright's treatment, since it is the simpler and better known.

For Wright, what is peculiar about teleological or functional statements is that they explain the existence of a current state in terms of its consequences. Thus, for Wright: "The function of X is Z *means* (a) X is there because it does Z, (b) Z is a consequence of X's being there."[57] This appears anomalous from the perspective of modern science since it seems to reverse the order of efficient causation. But if X does not specify a determinate individual, but rather any member of a class that is capable of replicating itself, then the anomoly is removed. What the account does is explain the occurrence of the present X as resulting from the fact that a previous X did Z, which made that X more adapted and so more likely to reproduce.[58] Hence, from Wright's perspective, it is the adaptedness of a previous X that explains the current X. Within this framework, one can say that something serves a particular function or purpose if it was selected and, hence, now exists because it met that need.

Some scholars have objected to making such use of evolutionary theory to give a naturalistic grounding to teleology. Margolis considers the possibility of sudden changes through which a new substance comes to play the same role as a current one (his example involves the rather unlikely event of CO replacing molecular oxygen in the atmosphere and also coming to be used by animals in their production of energy).[48] In such an instance, it makes sense to say the new substance is serving a function, but it does not according to this analysis because it was not its previous occurrence that gave rise to the current occurrence. This is, indeed, a serious objection. What it shows is that, although the connection between teleology and evolution may be correct, the connection is being made in the wrong order. The correct order is to claim that those things that are functional will evolve, rather than to claim that those things that evolved are functional.[59]

Reversing the connection between selection and teleology, however, imposes a new task. In Wright's account we could use selection to tell which things were functional; now we need an independent way of determining what is functional. We can do this by treating as functional those traits that promote or could promote the entity's selection in the current context. To identify such traits, one ultimately needs an engineering analysis that shows in detail how a trait meets selectional demands and so enhances the organism's capacity for reproduction. This, however, is something needed anyway if one is to avoid Gould and Lewontin's objection that evolutionary scenarios often amount to nothing more than "just so stories"; that is, stories simply made up to fit what is known about the course of evolution without having any independent confirmation.[60] Although such engineering analyses are available in only a few instances (*see* Williams, for a discussion of a few examples),[61] it is one of the tasks to which evolutionary biology is devoted. There is, however, a conceptually intermediate position that has been developed in philosophical reflections on evolutionary theory—the propensity interpretation of fitness.[62–64] What the propensity interpretation of fitness does is define fitness in terms of propensity to reproduce, not reproductive success itself. This is all that is required for our purposes, for we can now define something as functional if it increases the propensity of its bearer to reproduce. Using this notion, we can reverse the connection between selection and function as offered in Wright's account and so answer Margolis' objection.[65]

Thus far I have tried to show how an evolutionary scenario is able to ground a teleological perspective in biology. It is this teleological perspective that provides the necessary supplement to the concept of health based on the physiological concept of homeostasis that I began to develop earlier. The problem with that physiological perspective when taken on its own is that it failed to show why any premium should be placed on maintaining a system at its homeostatic equilibrium point. Within an evolutionary scenario, we can recognize the homeostatic design with its set equilibrium point as both a product of natural selection and, more important, as likely to enhance future survival. Maintaining the system in such a condition becomes important in aiding the survival and reproductive success of an organism (later I will indicate how the concept of reproductive success can be expanded to include more than biological reproduction). Not only does an evolutionary framework supplement the physiological concept based on the notion of homeostasis, it also indicates a way in which it may need to be modified. Sometimes the survival and reproductive abilities of an organism can be improved by altering the physiological constitution, including homeostatic design, that past evolution has provided. In such cases, altering the physiological constitution can be construed as improving the health of the organism.

Engelhardt has raised a pair of additional objections to using evolutionary criteria in evaluating what constitutes proper functioning and health. One objection is that what is beneficial in promoting the survival of the species may not be in the individual's interest: "Unlike evolution, which is a group- or species-centered concept, definitions of human health and disease tend to be individually oriented. Such definitions may not, therefore, include certain individual problems as states of health, even if those problems contribute to the survival advantage of the species."[2] Although Engelhardt does not give examples, he seems to be claiming that an evolutionary perspective cannot treat conditions that are detrimental to the individual but advantageous to the species as diseases. Engelhardt's objection, however, feeds on a misconception of contemporary evolutionary theory. He treats evolution as concerned with the promotion of the species, whereas the dominant tradition in evolutionary thinking has been to view selection as promoting individuals.[66],[67],[68] For the most part, group selection has been the object of derision, and some evolutionists have gone so far in the opposite direction as to insist that the proper focus of an evolutionary analysis is the single gene.[69] Only recently have a few biologists argued for viewing selection as working at the group or species level, as well as at the individual level, but they certainly have represented a minority perspective.[70],[71] Moreover, all philosophers and biologists arguing for higher levels of selection also grant that selection occurs at lower levels, such as the level of the individual. Later I will try to put to use the multiple perspectives one can take on selection, but for now it is sufficient to note that Engelhardt's analysis rests on a misunderstanding of evolutionary theory.[72]

Engelhardt, however, raises one additional objection that may more directly get at the heart of what seems troubling with an attempt to give a naturalistic interpretation of function in terms of evolutionary theory: such an explication "is to forego mention of why one might be interested in certain types of well-being (except for the commitments to the survival of the species)."[15] This challenge, however, may not be directed so much at this definition of health as toward the value of health, so defined. Although it is easy to see why many people may prefer health to the lack of health (it allows them to succeed within their biological or social niche), one can also create a scenario wherein someone may reject health. For example, one may find a particular unhealthy state, or an activity (such as smoking) that is likely to destroy health, to be pleasurable and so see no reason to avoid it. But, as the example of smoking should show, the fact that some people do not value health when so defined does not constitute an indictment of the definition of health.

Even though these responses deal with the specifics of Engelhardt's objections, there is an objection closely related to them that is more serious and requires a considerable broadening of the evolutionary framework employed so far. I will use an illustrative example to raise the objection. Conditions that afflict one after his or her reproductive years would not seem to reduce biological adaptiveness (propensity to reproduce) and so, within the evolutionary framework, could not count as diseases. There are two answers to this specific objection; one that stays within a narrow evolutionary framework and is in the spirit of sociobiology, and another that radically expands the evolutionary perspective. The first approach attempts to show that the ability of humans to perform certain activities after reproducing does help ensure the survival of their offspring. Thus, it increases their propensity of successful reproduction.

The applicability of this narrow answer to modern culture is dubious, since much of one's activity after reproducing (if one even reproduces) is not directed toward advancing one's progeny. Moreover, activities earlier in life are not all directed toward reproduction. The focus on reproduction in modern evolutionary theory seems ill suited to concerns about human health. However, the evolutionary framework need not be restricted to the biological level that focuses on individual reproductive success. By applying it more generally, one can perhaps find a better answer to this objection. Lewontin provides the basis for a more general application of the evolutionary perspective when he shows that anything can be understood as evolving as long as it has mechanisms for variation, selection, and retention.[73] Thus, any human products—ideas, cultural patterns, values—that can be selectively propogated can be construed as evolving (*see* Boyd and Richerson, who are adapting the evolutionary framework to show how aspects of culture evolve).[74] Many more of the characteristics of humans could be construed as adaptive when one extends one's focus beyond the inheritance of genes to the inheritance of cultural entities. Within this broader evolutionary framework, health involves those characteristics that make one adaptive in one's endeavors to propogate oneself or one's culture, and so on, when judged against a variety of selection forces, and disease is whatever deters from such adaptedness.

Employing such a generalized concept of evolution in a definition is clearly not without problems. Attempts to apply a generalized concept of evolution to other domains than biology have been severely criticized. Numerous philosophers have been outraged by Popper's, Campbell's, and Toulmin's proposals for an evolutionary epistemology (according to which the growth of knowledge results from a process of blind variation and selective retention); an evolutionary account of the development of ethical standards would undoubtedly receive an even worse reception.[75-77] What these philosophers resist is the loss of the purely normative side of these endeavors as practiced within philosophy (knowledge and moral value become equated with what people accept). Numerous anthropologists, suspecting that another version of sociobiology is being offered, have rejected Boyd's and Richerson's schema for the evolution of cultural ideals.[74] Moreover, employing such a generalized concept of evolution in the endeavor to define health may open Pandora's box. One can view almost any human product as capable of selective replication (including, for example, pornography). If health is whatever promotes generalized evolutionary success, it must be directed toward promoting all of these modes of replication. One of the motives that led Kass to seek a naturalized concept of health was to constrain the domain of health, since he was concerned about the temptation of health-care workers to extend their purview to almost all aspects of human activity. This proposal seems to lack any similar constraint. Finally, it is not at all clear that a generalized concept of evolution is consistent in that what may enhance human performance against one of these sets of selection criteria may inhibit it with regard to others. This would force a judgment as to which selection forces governing human endeavors it is most important to satisfy. If this judgment ultimately has to be ideological in character, then one has foresaken a naturalistic framework. It may be, however, that biological evolution could give some guidance for evaluating which sets of selection forces are compatible with human survival, and that this would provide enough constraint on which of these selection pressures it is acceptable to attempt to satisfy.

Although admitting that these problems confront the concept of health based on a generalized evolutionary framework that I am developing, I will not deal with them further here. Rather, I will focus on some of the strengths of such a concept, recognizing that there are further difficulties to be resolved. Comparing it first with the naturalistic approaches considered above, this approach offers a framework for evaluating when an organism is functioning properly, something the others could not do in a nonquestion-begging way. The approach I am exploring directs one to engage in an engineering analysis to identify how the physiological organization of the system equips it to deal with the selection forces working upon it. A healthy state of the system is one in which it makes best use of its physiological endowments in responding to selection pressures. Moreover, this approach is able to meet one of the serious objections raised against earlier naturalistic attempts to define health; namely, that what constitutes health varies depending on the environment. Within the evolutionary framework, the evironmental relativity insisted on in the objection is incorporated directly into the concept of health, since the selection forces operating in the individual's environment play a central role in determining what constitutes his or her healthy state.[78]

In addition to answering the major objections to earlier naturalistic concepts of health, the position I have outlined here offers an advantage over the relativistic positions of Engelhardt and Margolis. I noted earlier that Englehardt's conception, according to which the notion of health depends on the values of one's society, is not in a position to condemn the abuses of human beings in the name of health that occurred in the United States in the era of slavery, in Germany during the Holocaust, and today with psychiatry in the Soviet Union. The naturalism of the position I am proposing provides a foundation for saying that these abuses are incorrect uses of health care, for they do not advance the adaptiveness, biological or social, of the individuals who are the recipients of such care. They represent the imposition of ideology on patients, not an endeavor to equip patients to meet the demands of their environmental niche. By insisting that health care promote the adaptiveness of patients, we escape the relativism of Engelhardt's position. Boorse also tried to avoid this relativism, but at the expense of limiting health to those conditions needed within any ideological framework, a domain that might turn out to be empty. The concept of health I am offering is not so limited, for it allows within the domain of health those features of human life that we need for meeting whatever selection forces we confront.

Finally, insofar as this concept focuses on health rather than disease, it can answer the objections raised earlier to the ontological concept of health that make disease the primary notion. I have already indicated that within this framework one expects a many–many mapping between causes and actual disease states, so that this does not provide an obstacle as it did for one version of the ontological concept. The more general objection to ontological concepts was that there seems to be nothing in common between different diseases that makes them all diseases. If health is taken as the primary notion, though, there is no need to find a common characteristic that makes states diseased. Rather, disease is simply a disruption from a state of proper functioning.

In this section I have articulated a concept of health that is based on a physiological framework embedded in an evolutionary model. This concept is naturalistic in character, and we have seen that it overcomes objections to earlier naturalistic concepts and is also preferable to the nonnaturalistic alternatives. I have acknowledged that there are unresolved difficulties confronting this conception; in particular, difficulties in developing the generalized evolutionary framework to which I appealed. But the strengths of the concept here indicated provide reason to consider it further. One additional way to evaluate this model is by exploring its implications for various domains of health. In the following section I will do this for the domain of genetic health.

Application to the Case of Genetics

In turning to the domain of genetic health, my goals are modest. I will attempt only to establish that the implications for this domain of the physiological concept of health I have advanced are sufficiently plausible for it to merit further consideration. I will not attempt to show that it provides the definitively correct concept of genetic health. I have chosen the domain of genetic health for a number of reasons. First, it is an area in which the traditional physiological concept of health seems most problematic. In this domain we are confronting the issue of what the character of the species ought to be, whereas the only reference point for the traditional physiological concept is the actual character of the species. Hence, if the conception I have proposed suggests a coherent framework for discussing genetic health, that will constitute progress. Second, many contemporary health controversies involve genetics, so any insights the proposed concept of health can offer may be of use in public policy deliberations. However, this is also a dangerous area. One of the specters that seems to loom as soon as one introduces the notion of genetic health is that of eugenics—the movement to improve the human species by altering its gene pool. The eugenics movements are dangerous, in part because without the ability to objectively specify whether a condition is healthy or not, these movements degenerate into endeavors to promote one's own racial or ethnic characteristics. Although the potential for abuse will remain, a naturalistic physiological concept of genetic health may provide the basis for critiquing these abuses (something that nonnaturalistic concepts such as Englehardt's cannot do). This is the third reason for trying to apply it to this domain.

There is a further aspect of the social concern about eugenics that will also be a concern for this or any other concept of genetic health. Even if we had an objective foundation for evaluating genetic health, so that it was not simply a notion that could be put to ideological use, it could still have dangerous implications for a society insofar as it provided the basis for discriminatory treatment of those human beings judged to be genetically diseased. Labeling certain people "diseased" as a result of their very constitution prepares the way for discriminatory social policy. In particular, if the means one adopts to improve genetic health is to prohibit genetically diseased individuals from reproducing (this is a prominent, but clearly not the only, means one might employ), one is challenging what many individuals take to be a basic liberty. These dangers are heightened because genetic diseases are often found among minority groups.[79]

Despite these dangers, however, there are convincing reasons to focus on disease conditions that result from an individual's genetic constitution. We can trace conditions like Tay Sachs, sickle cell anemia, and Down's syndrome to the genetic constitution of the individual, and these conditions do, undeniably, resemble other diseases in the suffering and impairment of function they induce. To avoid the dangers in making the genetic constitution parts of the disease concept, one might try to construe only the manifestations of the genetic conditions as the disease. There is a theoretical basis to support this move. Phenotypic conditions result not from the genes alone, but from the interaction of genetic material and environment.[80] In the case of some genetically abnormal conditions (PKU deficiency), it is possible to treat the organism so as to overcome the consequences of the genetic condition. However, this is not a satisfactory resolution of the problem. If adequate theraputic regimens could be devised for all cases now considered genetic diseases, that would be the beginning of ever-increasing demands on the health-care establishment. If the genes responsible for these conditions are not selected against, they will remain in the gene pool and may even increase in frequency (a phenomenon known as increasing genetic load).

What these arguments suggest is that one should not try to avoid the task of defining genetic health and illness, but see if an objective, naturalistic concept of genetic health is possible. One will then have to investigate other procedures to deal with the other social dangers such a notion raises. This may not be easy, but it is not a problem unique to the domain of genetic health. Many other conditions, when they have been classified as unhealthy, have brought with them a stigma. One need only consider the stigma many people still attach to mental patients. Although not trying to belittle the significance of these social dangers in developing and employing the notions of genetic health and genetic disease, I will not deal further with them here, but will consider the utility for this domain of the physiological concept of health embedded in an evolutionary framework that I have developed.

This physiological concept would count one's genetic constitution as healthy if it equipped one to meet the biological and social selection forces operating in one's environment. One implication of this should be noted immediately: Genetic health is environment-relative. A gene, or even a whole genome, is not healthy because of its intrinsic character, but because of its ability, when placed in a particular environmental niche, to produce a phenotype well adapted to that environment. This, however, should not be surprising, since it is simply a reflection of the environmental relativity the evolutionary framework imposes on health assessments generally. However, one must be careful about the timeframe one employs in evaluating adaptedness. An organism that becomes too specialized as a result of adapting to the demands of its immediate environment may limit the survival capacities of its offspring who face a changed environment. Hence, in invoking this framework in dealing with human health, one must adopt a longer-term perspective, and evaluate genetic health not just in terms of whether the current organism will flourish in its current niche, but whether it will produce offspring capable of flourishing in projected future environments. So the concept of genetic health should be modified so that we would judge an organism to be genetically healthy if its genes produce a phenotype well adapted to its current niche and capable of producing offspring well adapted to the niches they will likely occupy.

It should be clear that this concept of genetic health drawn from the general framework I outlined earlier does not rest on a particular ideological concept of what an ideal human is, and so is not as prone to the kinds of abuse that have been fostered by the various eugenics movements of the recent past. However, it may seem to encounter other difficulties that troubled these programs as well. One of these stem from the fact that the notion of genetic health itself challenges our concept of human value by offering an ideal that all current human beings may fail to satisfy. All of us suffer some deficits in dealing with the demands of our environment that can, at least in part, be traced to our genes. Moreover, it seems to open the possibility of radically altering human nature to create humans that conform to these standards. Finally, the notion of being equipped to meet the selection forces in one's environment is typically a relative notion: One organism is judged better equipped than another. This makes health a competitive matter.

Explaining how the notion of genetic health advanced here can deal with these difficulties will help to clarify its character. In response to the first difficulty, note first that there will not be one ideal of genetic health, since the concept is environment-relative. But even more generally, there is not one genetic condition that is ideal for each environment. Those with different genetic endowments may find different ways of dealing with the demands of the environment, just as do different species that occupy the same environment. On the other hand, it is probably the case that each of us could deal with our environment better if we had some genetic alteration. However, given the complexity of the relationship between our genome and the environment, it is unlikely that we could be sure which alterations would make us better adapted. To use Simon's terminology, evolution follows a "satisficing," not an "optimizing" strategy—it seeks solutions that work adequately, not optimally, since there are enormous costs in trying to reach a better solution. For pragmatic reasons, we would do well to follow evolution in this and not impose optimal adaptedness as our standard for genetic health. Rather, we should construe genetic health in terms of those conditions that allow individuals to function relatively well in the environments they confront and construe genetic disease as those genetic conditions that impose obvious and severe impairments on the functional capacities of human beings. If genetic health and disease are construed this way, these notions do not involve advocating radical alterations in the human genome. If specific ways of improving the genome are discovered and we can be confident that they do not entail side effects that are detrimental, then these might be advocated in the name of genetic health. The antecedents of the previous conditional statement, however, clearly impose tough standards.

The final difficulty noted above is that by introducing the evolutionary framework into the concept of genetic health, one seems to make genetic health a competitive matter. Evolution has, indeed, often been portrayed as a competitive process, with each organism having as a goal to leave more successors than others. Not all aspects of evolution are competitive—the focus on competition is perhaps only a remnant of the Victorian heritage of Darwinian evolutionary theory. Occasionally organisms are in competition for the same niche, but more often each organism inhabits its own niche and tries to meet the challenges offered by that niche. In these circumstances, success for an organism can be viewed as maintaining itself over generations in that niche, or some successor niche. Surviving competition for the niche is only one of the requirements for maintaining a niche for oneself. If this perspective on evolution is employed, one does not turn the quest for genetic health into a competitive affair.

So far I have looked at genetic health from the perspective of the individual. However, one can also move to a higher level and consider the evolution of a group or species. When one does so, one finds a perspective that even more sharply differentiates the perspective of genetic health proposed here from that advocated by most eugenics programs. Eugenics movements have typically operated with the idea that there is a genetically most fit condition that could be advanced by a program of controlled breeding. I have already noted one way this idea is biologically wrong: Organisms with different genetic endowments can find different ways of meeting the demands of their niche. But the idea is also incorrect when looked at from the societal level. If variety is eliminated from a species, it may become particularly well adapted to one niche, but at the cost of being extremely vulnerable to any change in the niche. In order for a group (family, species, and so on) to continue to leave offspring once the environment changes, variation within the population is essential.[81] Hence, a societal eugenics program directed toward promoting one human type amounts to a program for societal suicide.

The importance of variation for a population produces a paradox when considered from the perspective of an individual in the population. The individual needs variability in the gene pool of the population so that he or she will be able to produce a variety of offspring that can survive in different possible future environments. However, this variability can also be the source of genetic disease of offspring. Many of the most common genetic diseases result when both parents carry a recessive gene (a gene only expressed in the homozygotic state), and these come together in a homozygotic state in an offspring and so produce the harmful condition. Those individuals with greatest genetic variety are also the one's most likely to carry harmful alleles and produce offspring with genetic diseases. Thus, Lappe comments "Thus, the greatest paradox of the concept of genetic health is that while from a population point of view, the 'healthiest' individuals are probably those who carry the greatest number of genes in the heterozygous state, they are also the ones at greatest risk for transmitting genetic disease."[82]

To resolve this paradox, we ought not reject genetic variability, for we have already seen the harm that can produce. Rather, the focus should be on eliminating those alleles from the gene pool that are responsible for severe genetic disorders when brought into a homozygotic state. However, this raises another issue: Are persons unhealthy because they can transmit a gene for a disease, even if they do not suffer from the disease themselves? Since reproduction is a major biological goal, the potential to produce a handicapped offspring would seem to be a relevant factor in assessing how adapted one is. If this criterion is invoked, few if any people will turn out to be healthy, because almost all of us possess at least one gene that would be deleterious in the homozygotic condition. However, one can gain some perspective of this situation if one considers only those with whom an individual has a high likelihood of mating, and considers only one (or a couple of) generations. Then one can evaluate an individual's genetic adaptedness simply in terms of the risk of that individual producing a handicapped offspring with those with whom he or she is likely to mate. By adopting this perspective, the energy of those trying to advance genetic health will be directed to those recessive genes that, because of the way they are distributed in a population, are most likely to be expressed.

There is one further feature of the concept of genetic health being proposed that should be discussed. According to this concept, because the health of an organism is relative to an environment, then the organism is no longer healthy if the environment changes so that the organism is no longer well adapted. This makes it seem that the responsibility lies with the individual to be adapted to the environment. Most people would probably admit that individuals have some responsibility to adapt to their environment, but we are increasingly becoming aware that we have responsibilities to maintain the environment so that we can survive in it. Thus, we speak of the obligation to maintain a healthy environment. To accommodate this, though, one only needs to take seriously the notion within evolutionary theory that adaptation is a relational notion. It involves a relation between an organism and an environment, and so one can improve the adaptation by changing either relata. The tendency to focus on the organism and not the environment when something is not adapted is a manifestation of the danger of localizing diseases I identified in discussing the ontological conception of diseases. A physiological conception of health grounded in a evolutionary perspective, however, provides a means of rebalancing the focus and recognizing that adaptation involves having a proper match of organism and environment. Sometimes we should direct our concern for improving health to the environment and not to individuals.

In this section I have explored the implications the physiological concept of health grounded in an evolutionary perspective that was developed in the previous section has for the domain of genetic health. Although not considering all the implications of this framework for genetic health, I have tried to show that the concept that results is a viable one, and, in particular, is not subject to the same difficulties as the concepts advanced by various eugenics movements. However, here I have only begun to investigate the applications of this physiological concept of health. This preliminary inquiry shows that it has promise, but further investigation is required to see if that promise can be fulfilled.

Notes and References

[1]T. Parsons (1951) *The Social System*. Free Press, New York.

[2]H. T. Engelhardt, Jr. (1978) Health and Disease: Philosophical Perspectives in *Encyclopedia of Bioethics*. (W. T. Reich, ed.) MacMillan, New York.

[3]L. S. King (1982) *Medical Thinking: A Historical Preface*. Princeton University Press, Princeton.

[4]T. Szasz (1974) *The Myth of Mental Illness: Foundations of a Theory of Personal Conduct*. (Revised ed.) Harper and Row, New York.

[5]W. K. Goosens (1980) Values, health and medicine, *Phil. Sci.* 47, 100–115.

[6]R. Dubos (1959) *Mirage of Health: Utopias, Progress, and Biological Change*. Harper and Row, New York.

[7]R. Dubos (1968) *Man, Medicine, and Environment*. Frederick A. Praeger, New York.

[8]L. Kass (1975) Regarding the end of medicine and the pursuit of health. *The Public Interest* 40, 11–24.

[9]H. T. Engelhardt, Jr. (1984) Clinical Problems and the Concept of Disease, in *Health, Disease, and Causal Explanations in Medicine*. (L. Nordenfelt and B. I. B. Lindahl, eds.) Reidel, Dordrecht.

[10]King traces this view back to Sennart in the 17th century.[3] Kass, however, takes issue with this claim.[8] He points first to etymological considerations, showing that in English as well as in Greek, Latin, Hebrew, and German, the word for health has a distinct heritage from the terms of disease, illness, and sickness. It is unclear, however, what conclusion one should draw from such etymological considerations. For two terms to be complementary, it is not necessary that they were originally understood as complementary. Even if Kass is right and disease was first understood as an invasion of the body and health as a state of proper functioning of the person, it can be discovered later that it is when a person is invaded by disease that the person ceases to function well. This is comparable to the discovery that the morning star is the evening star. What would render this assessment wrong is if there were infringements on the proper functioning of the body that nonetheless did not involve disease. Under this circumstance, one might hold that health care should be concerned only with combating diseases, and not with other dimensions of promoting proper functioning, such as education. But although Kass is interested in restricting the domain of the health-care profession, he does not use this argument schema, for he also wants to restrict the concept of health and not apply it to all aspects of human functioning. Thus, the point of his denying the complementarity of the concepts "health" and "disease" is not clear. (Whitbeck, distinguishes the two concepts in a different manner, pointing to a special value-ladenness that arises with the concept disease.)[83]

[11]J. Rawls (1971) *A Theory of Justice*. Harvard University Press, Cambridge.

[12]F. J. V. Broussais (1824) *Examen des doctrines médicales et des systèmes de nosolgie*. Mequignon-Marvis, Paris.

[13]G. B. Risse (1978) Health and Disease: History of the Concepts, in *Encyclopedia of Bioethics*. (W. T. Reich, ed.) MacMillan, New York.

[14]H. T. Engelhardt, Jr. (1975) The Concepts of Health and Disease, in *Evaluation and Explanation in the Biomedical Sciences*. (H. T. Engelhardt, Jr. and Stuart Spicker, eds.) Reidel, Dordrecht.

[15]T. Sydenham (1676) *Observationes medicae circa morborum acutorum historiam et curationem.* Translation by R. G. Latham London. G. Kettilby, in *The Works of Thomas Syndenham.* 1848–1850.

[16]P. Pinel (1798) *Nosographie philosophique, ou la methode de l'analyze applique á la mèdecine.* Richard, Caille, Et Ravier, Paris.

[17]For Sydenham, "a disease . . . is nothing more than an effort of nature, who strives with might and main to restore the health of the patient by the elimination of the morbific matter." Sydenham's version of the ontological conception is thus clearly the first version, which hypostatizes diseases as logical types, but does not identify them in terms of disease entities that are taken to be the causes of the symptoms.

[18]A. Benivieni (1507) *De abditis nonnullis ac mirandis morborum et sanationum causis* in *The Hidden Causes of Diseases.* Translation by Charles Singer, 1954. Charles C. Thomas, Springfield Illinois.

[19]G. Morgagni (1761) *De Sedibus et causis morborum pes anatomen indagatis.* Venice.

[20]X. Bichat (1801) *Anatomie générale applique à la physiologie etàla médicine.* Broson, Paris.

[21]The ability to identify lesions in living patients as well as cadavers was made possible by Auenbrugger's discovery that sound elicited by percussion of the chest could provide information on anatomical changes occurring there, and Laennec's introduction of the stethoscope as a tool for detecting these sounds. (*see* Temkin, for further details).[84]

[22]G. L. Engel (1960) A unified concept of health and disease. *Pers. Biol. Med.* 3, 459–485.

[23]G. Fracastoro (1546) *De Contagione et Contagiosis Morbis et Eorum Curatione.* Translation by W. C. Wright. G. P. Putnam, New York.

[24]R. Koch (1884) The etiology of tuberculosis. *Reichsgesundheitsamt Mitteilungen 2* 1–88. Translated and reprinted in H. A. Lechevalier and M. Solotovosky, eds., *Three Centuries of Microbiology.* Dover, New York. 1974.

[25]C. Eijkman (1897) Eine beriberiahnlicke Krankheit der Huhner. *Virchow's Arch.* 148, 523.

[26]G. Grijns (1901) Over Polyneuritis gallinarum, *Geneeskundig tijdschrift voor Nederlandsch-Indie* 41, 3. Portion translated in T. S. Hall, *A Source Book in Animal Biology.* Harvard University Press, Cambridge.

[27]F. G. Hopkins (1906) The analyst and the medical man. *The Analyst* 31, 385–404.

[28]F. G. Hopkins (1912) Feeding experiments illustrating the importance of accessory factors in normal dietaries, *J. Physiol.* 44, 425–460.

[29]C. Funk (1922) The etiology of the deficiency diseases. *Jo. State Med.* 20, 341–364.

[30]W. Bechtel (1984) Reconceptualizations and interfield connections: The discovery of the link between vitamins and coenzymes. *Phil. Sc.* 51, 265–292.

[31]The localization spoken of here does not have to be physical localization.

Assigning functions a place in a flowchart counts equally as localization as I am using the term.

[32]W. Bechtel (1982) Two common errors in explaining biological and psychological phenomena. *Phil. Sci.* 49, 549–574.

[33]W. Bechtel, (1982) Functional Localization Assumptions as Impediments to Clinical Research in *A General Survey of Systems Methodology*. (L. R. Troncale, ed.) Vol. II. 953–962. Intersystems Publications, Monterey, California.

[34]W. C. Wimsatt (1980) Reductionistic Research Strategies and Their Biases in the Units of Selection Controversy in *Scientific Discovery: Case Studies*. (T. Nickles, ed.) Reidel, Dordrecht.

[35]H. T. Engelhardt, Jr. and E. L. Erde (1980) Philosophy of Medicine in *A Guide to The Culture of Science, Technology, and Medicine*. (P. T. Durbin, ed.) Free Press, New York.

[36]G. L. Spaeth and G. W. Barber (1980) Homocystinuria and the passing of the one gene–one enzyme concept of disease. *Jo. Med. Philos.* 5, 8–21.

[37]E. Rosch, C. B. Mervis, W. Gray, D. Johnson, and P. Boyes-Braem (1976) Basic objects in natural categories. *Cognitive Psychol.* 3, 382–439.

[38]Engelhardt offers an interesting suggestion.[9] He proposes that for clinical applications, one start not with the notions of "health" and "disease," but with "clinical problems" that patients present. According to this proposal, the common characteristic of clinical problems would be the fact that they are what patients present and for which they seek treatment. This may, in fact, be the operating criterion in normal practice, but it poses the problem that there may be no limit to what could fall within the domain of health-care workers.

[39]O. Temkin (1963) The Scientific Approach to Disease: Specific Entity and Individual Sickness in *Scientific Change: Historical Studies in the Intellectual, Social, and Technical Conditions for Scientific Discovery and Technical Invention, from Antiquity to the Present*. (A. C. Crombie, ed.) Basic Books, New York.

[40]To this Boerhaave added the complementary definition of disease in terms of the interruption of functions: "The condition of the living body whereby the ability of exercising any one function is abolished, is called disease."[39]

[41]Temkin notes this teleological feature of the physiological attempts to define health and disease, and presents the problem it raises: "The answer (as to whether disease can be defined in terms of departure from normal function) may depend on the biologist's philosophical orientation. If he [or she] excludes all teleology, he [or she] may . . . believe that health and disease are not scientific terms. If he [or she] does not mind using teleological notions, he [or she] may find it pertinent to pay attention to those states where nature fails in two of its main aims, as Galen had them, assurance of the life of the individual and of the species."[39] Later I will argue that the matter is more than one of predilection by showing that there is an objective basis for insisting on a teleological element within biology.

[42]C. Boorse (1975) On the distinction between disease and illness. *Philos. Public Affairs* 5, 49–68.

[43]E. Mayr (1976) Typological Versus Population Thinking in *Evolution and*

the Diversity of Life. (E. Mayr, ed.) Harvard University Press, Cambridge, Massachusetts.

[44]C. D. King (1945) The meaning of normal. *Yale Journal of Biology and Medicine* 17, 493–494.

[45]The difficulty that exists here has sometimes been obscured by a confusion over two senses of normal, one in which it refers to the mean condition, the other in which it refers to the optimal condition. The two are clearly not equivalent. When the development of hospitals in the 18th century made possible the examination of large numbers of individuals, it was thought to be possible to determine a set of norms for the healthy person. For example, the discovery of the thermometer gave rise to the suggestion that there is a normal body temperature for humans, such that deviations from this norm could be taken as a sign of disease. In some cases, the normal condition, in the sense of average may also constitute the correct normative condition, the healthy condition. But one can readily acknowledge circumstances in which the prevalent mean would not constitute a health state (e.g., during an attack of the plague). One needs to be careful not to equivocate between these two senses of "normal." *See* Grene, who distinguishes not only these two senses but a third as well: being "characteristic for members of the species."[85]

[46]Macklin laments the lack of an integrated theoretical perspective, suggesting that if there were such, it would offer the foundations for the concepts of health and disease: "There is no general, well-integrated theory of the sort that exists in, say, physics, interconnecting the well-developed fields in medicine of physiology, anatomy, pathology, neurology, immunology, etc. with current developments in the biological sciences . . . [T]he absence of well-confirmed fundamental laws, from which other laws are derivable, and the absence of a systematic, general theory result in the situation that within medicine itself, there are no clear or precise formulations of the basic concepts of health and disease, and no set of necessary and sufficient conditions for their application."[86]

[47]I will focus on those who appeal to a society's values. Two philosophers who have made the individual's values central in defining health are Whitbeck and Porn.[83],[87] For Porn, health consists in an equilibrium between and individual's goals and his or her repertoire of abilities. This makes one unhealthy if one's goals are unobtainable. Nordenfelt suggests that Whitbeck can avoid this problem because in her otherwise very similar treatment, she introduces the notion of "real goals," but, as he notes, this raises the question as to how these are to be measured.[88] In particular, Nordenfelt suggests that one lands in a circularity when goals are interpreted as needs, for these needs are in turn identified as what is required to maintain health. However, contra Nordenfelt, it does not appear that Whitbeck's introduction of "real goals" was meant to eliminate unobtainable goals, but only to allow us, when we think someone's goals are irrational, to look for a more ultimate goal. It seems then, that for both Whitbeck and Porn, that one is ill if one has set unobtainable goals, which certainly seems counterintuitive. Additionally, their approach permits no limits on the domain of health.

[48]J. Margolis (1976) The concept of Disease. *Jo. Med. Philos.* 1, 238–255.

[49]In addition to pointing out that standards are implicit in any judgment, Margolis offers three additional reasons for maintaining that these norms cannot, in any direct way, be "read off" the physical structure of the body: (1) there is not a one-to-one but only a many–many relation between disease condition and localized deficits, (2) many of the processes making for health are molar and not found in localized parts, and (3) the healthy condition of body parts can vary through ranges, and the point in the range that counts as healthy may depend on the current environment the individual is in. The last of these points is the same as that discussed previously and, as I indicated, I will try to make a virtue out of this purported difficulty. The first two do not seem to be particularly telling, since they are objections against a localized conception of a disease entity, not against the view of the defender of the physiological concept that one can determine objectively the health condition of an organism.

[50]T. Lenoir (1982) *The Strategy of Life*. D. Reidel, Dordrecht, Holland.

[51]C. Bernard (1865) *Introduction à l'etude de la médecine experimentale*. Bailliere, Paris. English translation by M. C. Greene, *An Introduction to the Study of Experimental Medicine*. Dover, New York 1959.

[52]C. Bernard (1859) *Leçons Sur les Propriétés Physiologiques et les Alterations Pathologiques des Liquides de L'organisme*. J. B. Baillière, Paris.

[53]In terms of this organization, Bernard proposed to account for some of the factors that had led vitalists to posit a vital force. For example, Bichat, a significant early 19th century vitalists, had considered two features of living systems to be indicative of the vital force active within them: their ability to respond to and restore themselves in the face of external stimuli and their indeterminacy in responding to stimuli (expressed by their failure to always respond to the same stimulus in the same way). Bernard showed that both of these features could be explained in terms of the organization found in the internal environment. The response to external stimuli was mediated by compensating mechanisms through which the organism attempted to restore its internal environment to a constant state. The variation in response to similar stimuli resulted from somewhat different internal conditions prevailing at different times (the differences resulting from different compensating mechanisms have been invoked in response to previous stimuli).[89],[90],[91]

[54]W. B. Cannon (1932) *The Wisdom of the Body*. Norton, New York.

[55]Dubos has made a similar point in a different manner: "In the light of these facts it is easier to understand why direct cause–effect relationships often fail to account for the natural phenomena of disease. Each type of insult—microbial invasion, chemical damage, physiological stimulus, or psychic event—can have many different effects depending upon the state of the recipient individual. On the other hand, any given pathological effect can be the outcome of many varied kinds of insults."[6]

[56]W. C. Wimsatt (1972) Teleology and the logical structure of function statements, *Studies in the History and Philosophy of Science* 3, 1–80.

[57]L. Wright (1976) *Teleological Explanations: An Etiological Analysis of Goals and Functions*. University of California Press, Berkeley.

[58]Evolutionary biologists since Darwin have used the term "fitness" ambiguously to refer both to the reproductive success of an entity and to its relation to the environment that makes for that success. Brandon has proposed employing the alternative term "adaptedness" for that which makes for the increased selection of the entity.[62] One of the benefits of this terminology is that it reveals that the principle of natural selection is not tautologous.

[59]Wimsatt seems to want to allow his functional analyses to interpret the relation between selection and teleology in either order.[56] He supplies the following general "normal form" for function statements: "According to theory *T* a function of behaviour *B* of item *i* in system *S* in environment *E* relative to purpose *P* is to do *C*." The purpose value in this statement is to be specified, at least in biological cases, by invoking evolutionary theory. Wimsatt goes on to discuss the evaluative and explanatory role of function statements, and notes that there are cases in which we want to invoke the function statement evaluatively, but not explanatorily. These are cases in which the behavior occurs in the current context independently of selection, but is now being selected for a certain purpose. In these cases the focus is on currently operating selection forces, but Wimsatt wants to claim also that functional statements lead us to try to explain the current occurrence of the entity as the product of past selection. The explanatory and evaluative endeavors are really much more distinct than Wimsatt indicates: One uses previous selection forces to specify the purpose variable, the other uses current selection forces.

[60]S. J. Gould and R. C. Lewontin (1979) The spandrels of San Marco and the Panglossian paradigm: A critique of the adaptationist programme, *Proc. Roy. Soc. Lond.* B205, 581–598.

[61]M. Williams (1982) The importance of prediction testing in evolutionary biology. *Erkenntnis* **17,** 291–306.

[62]R. N. Brandon (1978) Adaptation and evolutionary theory. *Studies in the History and Philosophy of Science* 9, 181–206.

[63]S. K. Mills and J. Beatty (1979) The propensity interpretation of fitness. *Philos. Sci.* 46, 263–286.

[64]R. N. Brandon and J. Beatty (1984) The propensity interpretation of 'fitness'—no interpretation is no substitute. *Philos. Sci.* 51, 342–347.

[65]Another objection of Margolis' seems simply to be erroneous. He claims: "one only has to imagine the evolution of diseased populations to see that creatures and organs and processes may well be said to function in a certain characteristic way, although on some relevant theory, we should not wish to say that the function of what is in question would then be given by the formulation cited."[48] What Margolis does not make clear is whether the "disease" is what enhanced the adaptedness of the organism and thus, provided for its selection. If not, then Wright's formula has not been satisfied and Margolis' example does not count against his proposal. If it does, then it is not clear why it is considered a disease. The one possible situation in which it might be considered a disease is when it once enhanced the adaptedness of the organism and is now a detriment (like sickle cell trait). This, however, can be easily

handled by the treatment proposed here that reverses the relation between evolution and functions. Within this account, sickle cell trait would count as healthy within an environment of malaria infestation prior to drug treatments, but subsequently to count as a disease.

[66]C. Darwin (1859) *On the Origin of Species by Means of Natural Selection*. Murray, London.

[67]T. Dobzhansky (1937) *Genetics and the Origin of Species*. Columbia University Press, New York.

[68]G. C. Williams (1966) *Adaptation and Natural Selection*. Princeton University Press, Princeton.

[69]R. Dawkins (1976) *The Selfish Gene*. Oxford University Press, Oxford.

[70]S. M. Stanley (1979) *Macroevolution: Pattern and Process*. W. H. Freeman, San Francisco.

[71]D. S. Wilson (1983) The group selection controversy: History and current status. *Ann. Rev. Ecol. Systematics* 14, 159–187.

[72]In a recent paper, there is a suggestion that Engelhardt may have revised his position, for there he alludes to the view that evolution selects for inclusive fitness, not individual fitness.[9] However, he seems to equate inclusive fitness with group fitness, which is only correct for certain special kinds of groups. Moreover, it is not clear that inclusive fitness is not one of the relevant frameworks for assessing health. What many of the sociobiologists who have invoked the notion of inclusive fitness have tried to show is that it could ground some of our social instincts. It may be that our abilities to fulfill these social functions is a relevant criterion in evaluating health.

[73]R. C. Lewontin (1970) The units of selection. *Ann. Rev. Ecol. Systematics* 1, 1–18.

[74]R. Boyd and P. Richerson (1985) *Conceptual Evolution*. University of Chicago Press, Chicago.

[75]K. R. Popper (1975) The Rationality of Scientific Revolutions in *Problems of Scientific Revolution: Progress and Obstacles to Progress in Science* (R. Harre, ed.) Claredon University Press, Oxford.

[76]D. T. Campbell (1974) Evolutionary Epistemology in (P. A. Schilpp, ed.) *The Philosophy of Karl Popper* (Vol. 1) Open Court, LaSalle, Illinois.

[77]S. Toulmin (1972) *Human Understanding: The Collective Use and Understanding of Concepts*. Princeton University Press, Princeton.

[78]In this respect, as in many others, the conception of health I am offering is quite similar to that of Dubos. According to Dubos, health is a matter of being properly adapted to one's environment. Although he does not emphasize the link with the evolutionary perspective, the notion of adaptation he is using received its natural interpretation in an evolutionary framework. Dubos appeals to what is fundamentally an evolutionary concept of adaptiveness to explain why the conditions making for health may be different in different circumstances—different conditions may be needed to make one adapted to one's environment. Moreover, Dubos stresses that within this framework one can see why the conditions that count as healthy cannot be permanently fixed. As one's environment changes, different conditions are needed to make one adapted.

In addition to showing how to account for the environmental relativity of health, Dubos raises an important caution as to the limitations of viewing health in terms of being adapted to one's environment: "The most disturbing aspect of the problem of adaptation is paradoxically that human beings *are* so adaptable. They can become adjusted to conditions and habits that will eventually destroy the values most characteristic of human life. If only for this reason, it is dangerous to apply to human beings the concept of adaptability in a pure biological sense."[7] Part of what underlies Dubos' concern can be addressed by employing the generalized evolutionary framework I sketched above. Another part depends on the distinction between long-term and short-term advantage. Biological selection will promote traits that offer a short term focus on the longer term and direct health care to promote long-term adaptation over mere short-term adaptation. This is not something that is possible for most other organisms, but it is within the potential of humans who have the ability to evaluate courses of action in terms of their anticipated long-term consequences.

[79]Confinement of genetics illness to distinct groups is both observed (consider Tay Sachs, thalosemia, and sickle cell anemia) and theoretically predicted. It is theoretically predicted since interbreeding is common among minorities excluded from the mainstream of the population, whereas interbreeding is common in the majority population. Inbreeding among a population in which a recessive gene is already in higher frequency because of past inbreeding results in the more frequent expression of recessive genes. (If one focuses on the genes, rather than the conditions produced by the genes, the evidence suggests that all humans, on average, carry five recessive genes that, expressed would produce the manifestations of genetic disease. However, when the breeding population is large, these recessive genes are almost never expressed.)

[80]R. C. Lewontin (1974) The analysis of variance and the analysis of cause. *Ame. Jo. Hum. Gene.* 26, 400–411.

[81]"There is mounting evidence that previous determinations of the 'cost' of genetic mutations were assessed too severely. One of the emerging concepts of genetics is that genetic diversity per se, where many individuals in a population carry a high proportion of genetic variants, is itself of great value for species survival. Such variability can assure the constancy of embryonic development in the face of environmental stresses, and can proffer a certain guarantee of flexibility in the face of environmental change. While there is no way for *everyone* in a freely marrying population such as ours to become, or remain, heterozygous for any one gene, the survival of the population is reinforced by many people being heterozygous for different genes—hence the imponderable difficulty of defining just which particular heterozygous state is desirable or undesirable."[82]

[82]Marc Lappe (1973) Genetic knowledge and the concept of health, *Hastings Center Report 3* 4, 1–3.

[83]C. Whitbeck (1981) A Theory of Health in *Concepts of Health and Disease* (A. L. Caplan, H. T. Engelhardt, and J. J. McCartney, eds.) Addison-Wesley, Reading, Massachusetts.

[84]O. Temkin (1973) Health and Disease in *Dictionary of the History of Ideas* (P. P. Wiener, ed.) Charles Scribner, New York.

[85]M. Grene (1976) Philosophy of Medicine: Prolegomena to a Philosophy of Science in *PSA 1976*. (Vol. 2) (P. Asquith and F. Suppe, eds.) Philosophy of Science Association, East Lansing, Michigan.

[86]R. Macklin (1972) Mental health and mental illness: Some problems of definition and concept formation, *Philos. Sci.* 39, 341–365.

[87]I. Porn (1984) An Equilibrium Model of Health in *Health, Disease, and Causal Explanation in Medicine*. (L. Nordenfelt and B. I. B. Lindahl, eds.) Reidel, Dordrecht.

[88]L. Nordenfelt (1984) On the Circle of Health in *Health, Disease, and Causal Explanations in Medicine*. (L. Nordenfelt and B. I. B. Lindahl, eds.) Reidel, Dordrecht.

[89]W. Bechtel (1982) Vitalism and dualism: Toward a more adequate materialism, *Nature and System* 4, 23–43.

[90]W. Bechtel (1983) Teleomechanism and the strategy of life, *Nature and System* 5, 181–187.

[91]W. Bechtel (1984) The evolution of our understanding of the cell: A study in the dynamics of scientific progress. *Stud. Hist. Philos. Sci.* 15, 309–356.

Health and Aging

A Literary Perspective

Suzanne Poirier

If health could simply be defined as freedom from disease, there would be no need for this collection of essays. That we are not satisfied with clinical definitions of health that refer to pathologial analysis, physiological definitions that consider systemic homeostasis, or verifiable accountings that set physical norms, suggests that there remains an unclassifiable—unquantifiable—dimension to the concept of health that is as variable and eccentric as human nature itself. Is a person with a clinically incurable, yet controllable, "disease" permanently unhealthy? Is a person with a physical disability unhealthy because he or she is not physically "normal?" Or, as will be considered in this paper, do we label people "sick" when their life cycle brings them to a stage of development in which they can expect their bodies *naturally* to function with more difficulty or at reduced capability?

Such questions raise three further questions: (1) How do people in the situations described above define themselves in terms of health?, (2) How do personal definitions of one's own health differ from others' perceptions and definitions of one's own health?, and (3) What are the possible consequences of these different ways of perceiving health?

Turning to literature for a definition of health is tantamount to looking for a straw in a haystack, rather than the proverbial needle: The problem is not where to find the straw, but, rather, which one to choose. Even within a body of literature limited to a particular period in history, cultural identity, or political ideology, writers' personal responses to illness remain highly individualistic. Literature itself poses a problem in defining health in its use of, in literary terminology, "voice." In Barbara Macdonald's essay, discussed below, "Do You Remember Me?," the voice of the speaker is Macdonald herself; the essay explores her own experience with growing old. When Theodore Roethke and Margaret Laurence, both in their middle age, created, respectively, "Meditations of an Old Woman" and 90-year-old Hagar Shipley, there was a deliberate distancing from, and thus, a conscious manipulation of, a character's thoughts and actions. These personae might share some of the emotions or insights of their creators, but the younger writers also drew on their empathic observations or ideas about aging.

In short, in turning to a combination of actual and imagined voices to formulate a definition of health, the literary analyst draws on both the experience and speculation of the chosen writers. That definition is, in turn, influenced by the works the literary analyst chooses to consider. The conclusions drawn from the few pieces of literature studied below will have some validity in that they represent thinking or emotions that appear in several writers' works, but they will also represent only one way of responding to the personal experience of defining one's own health in old age.

And aging itself I see as simply "failing," a painful series of losses, an inevitable confrontation with the human condition.[1]

"Well, there's nothing exactly wrong, *organically,*" he says, pleased with this impressive word. "Doctor Corby just thinks you'd be better off with proper care and all."

"Marvin—what's wrong with me?"

"Nothing much, I guess," he mumbles. "You're getting on, that's all."[2]

The first of these two passages comes from a personal essay by Cynthia Rich, herself 41 years old. The second quotation is taken from the novel *The Stone Angel,* by Margaret Laurence. Marvin is the 65-year-old son of the novel's 90-year-old protagonist, Hagar Shipley. Both passages present an attitude toward aging as an unavoidable, "natural" illness. In the novel, however, Hagar is truly ill, suffering from some unnamed disease that eventually causes her death. That her son can avoid telling her about her condition in the name of old age—that he believes that simply dismissing her growing discomfort as "getting on" would be a perfectly acceptable explanation to Hagar herself—echoes Rich's sentiment that aging is "failing." Such a definition of old age by young (or at least younger) people places special burdens upon aging men and women.

One of the first steps in confronting old age is the awareness of one's physical changes. Hagar Shipley, even before she becomes truly ill, describes a body that can no longer respond as quickly as her mind. As she rises from her chair with her accustomed alacrity, her temporarily forgotten arthritis "knots inside [her] legs as though I had pieces of binder-twine instead of muscles and veins," and she "stumble[s] a very little over the edge of [her] bedroom rug." She is annoyed with herself for the unnecessary awkwardness and the debility it suggests, and she grumbles to herself, "I could be all right—I could right myself—if only she [the daughter-in-law] would not take alarm and startle me, the fool."[2]

Hagar's body has become, at times, awkward and cumbersome to her. It has not become her enemy, however. Doris more nearly plays that role, as she responds impatiently and blamefully to the changes that have occurred in Hagar's legs and feet. Hagar, in fact, has grown quite accommodating of the demands on her body. After being helped back onto her feet by Marvin, she shakes his "paw" from her elbow, and the three of them proceed once more out of the room: "I wait, summoning poise. . . . And then I venture down the stairs. I hold the banister tightly, and of course I'm all right, perfectly all right, as I always am when I haven't got an audience."[2] Hagar has had to learn new, slower, and more deliberate ways of getting about, but she is "all right, perfectly all right," as she has known all along. She has seen the changes in her body and has accommodated herself to those changes. Marvin and Doris appear to have seen the changes, but have responded instead with irritation and condescention.

In another example of encountering the physical consequences of one's advancing age, Barbara Macdonald, the 65-year-old woman whose essays form a dialog with those of Cynthia Rich, in *Look Me in the Eye: Old Women, Aging and, Ageism,* reports feeling, at times, quite detached from her old body:[1]

> Sometimes lately, holding my arms up reading in bed or lying with my arms clasped around my lover's neck, I see my arm with the skin hanging loosely from my forearm and cannot believe that it is really my own. It seems disconnected from me, it is someone else's, it is the arm of an old woman.[1]

The detachment from her own body, however, is not a reflection of repugnance to, or denial of, her changing skin. Rather, Macdonald expresses an almost heightened awareness of her body, similar to that of Laurence's Hagar Shipley preparing every muscle to walk down a flight of stairs. Macdonald senses a deliberateness about her physical self, and she watches it with some curiosity and interest. She continues:

> Standing before the mirror in the morning . . . I see that the skin hangs beneath my jaw, beneath my arm; my breasts are pulled low against my body; . . . [M]y body is being drawn into the earth—muscle, tendon, tissue and skin is being drawn down by the earth's pull back to the loam.[1]

Macdonald expresses her awareness of aging more metaphorically than does Hagar Shipley, but both women, one in life and one in fiction, find themeselves, as they grow old, rediscovering their bodies, relearning its ways, and adjusting their lives to its demands. Macdonald finds comfort in the idea of being drawn back to the earth. She finds it reassuring that, in a life that for her has "been filled with uncertainties," there will one day be one certain "time I can rely on myself, for life will keep her promise to me. I can trust her. . . . She will give me finally only one choice, one road, one sense of possibility."[1] Rather than coming to think positively about death, Hagar Shipley's awareness of her old age gives her a heightened awareness of life. "Each day," she muses, "so worthless really, has a rarity for me lately. I could put it in a vase and admire it, like the first dandelions, and we would forget their weediness and marvel that they were there at all."[2] Both women speak with calm joy of a heightened sensitivity to themselves and their world brought about by their aging. They face life more evenly, with a greater appreciation of its subtleties and nuances.

In his "Meditations of an Old Woman," Theodore Roethke makes similar observations. The old woman of his poems remarks that in her old age she has come to "prefer the still joy:/The wasp drinking at the edge of [her] cup;/A snake lifting its head;/A snail's music."[3] Roethke's old woman, however, reached this point only after contemplating the physical changes of aging, as do Hagar Shipley and Barbara Macdonald. The old woman of Roethke's poem echoes all three women cited so far when she speaks of aging at first as loss, decline, and barrenness:

> The spirit moves, but not always upward,
> While animals eat to the north,
> And the shale slides an inch in the talus,
> The bleak wind eats at the bleak plateau,
> And the sun brings joy to some.[3]

Roethke's old woman finds her body unfamiliar to her; she calls it "a strange piece of flesh." However, it often seems antipathetic to her conscious being: "the rind, often, hates the life within."[3] She likens her efforts to understand and accept these changes to a crab trying to walk along the bottom of a muddy pond. The effort is awkward and difficult, but nevertheless continues resolutely.

In the third meditation, called "Her Becoming," the woman begins to resolve her ambivalence about aging. She has accepted the quieter life, has "learned to sit quietly," and appreciates small sensations, such as the sight of wind ruffling the back feathers of birds in the gravel outside her window. But she also distinguishes between her body and her spirit:

> My shape a levity—Yes!—
> A mad hen in the far corner of the dark,
> Still taking delight in nakedness . . .
> In the back of my mind, running with the rolling water,
> My breast wild as the waves. . . .[3]

The persona lives a dual life, quiet and unremarkable—even comic, "a levity"—on the outside, passionate on the inside. In this way she is similar to Laurence's character, Hagar, whose outward unresponsiveness is often a mask for the fullness of her inner life. "Now I am rampant with memory," Hagar describes herself at one time.[2] Roethke's speaker is also like Barbara Macdonald, with her new closeness to the earth.

This dual existence is not achieved by dissociation from one's aged body. Instead, the persona's changed body is in some way necessary to her spiritual awakening. As she describes it, "From the folds of my skin, I sing."[3] She continues to puzzle about her changed body, asking at times "Do these bones live? Can I live with these bones?" [3] but her doubts are always overshadowed, overcome, by the special strengths her old age gives her:

> I take the liberties a short life permits—
> I seek my own meekness;
> I recover my tenderness by long looking.
> By midnight I love everything alive.[3]

All three women—one an actual person, one a fully drawn fictitious character, and one a poetic voice—encounter in old age the foreignness of their own bodies. At times that body is seen as separate from them, but ultimately all three women redesign their lives to accommodate their altered bodies. More important, they redesign their thinking about themselves and about the world around them. After initial questioning, they enter old age with poise, appreciation, and, at times, welcome. These women consider themselves healthy in terms of physical energy and ability, equal to the tasks they choose to undertake. That Hagar must descend the staircase more carefully than in years before does not mean that she cannot perform this task; she simply adjusts her expectations of how it will be done. That Roethke's old woman can spend hours sitting and watching the hens doesn't mean that she has mentally or emotionally deteriorated.

The concept of health as accommodation of change is a functional definition and, as such, is not a new idea. What makes it worth considering here is the conflict that arises when other people do not share the older person's more personal, less exacting definition of health. Roethke's persona states the problem the most vividly in her second meditation. She finds that her once familiar world has changed, has become more complex, and because of the new difficulties these changes bring with them, she herself may be changing:

—How needles and corners perplex me!
Dare I shrink to a hag,
The worst suprise a corner could have,
A witch who sleeps with her horse?
Some fates are worse.[3]

The persona at first characterizes herself in her increasingly restricted world with images that have uncomplimentary associations with old women—hags and witches. Old women, sometimes physically stooped, with wispy hair, their voices perhaps grown shriller or more brittle, often are given pejorative labels by the younger people who are dismayed or repelled by the physical facts of old age. But Roethke's old woman goes on to see old age with new eyes, knows that the condition that prompts the label might not be so bad after all, and concludes, "Some fates are worse."

The old woman of Roethke's poem does not go on to confront the labels and stereotypes by which younger people define old age as unhealthy. Hagar Shipley and Barbara Macdonald do. Hagar, as previously mentioned, is actually ill, but it is her general condition of old age that her family complains most about. She observes that Marvin has certain preconceptions about how old people should act. Sometimes she goes along with that preconception, "for the sake," she says "of such people as Marvin who is somehow comforted by the picture of old ladies feeding like docile rabbits on the lettuce leaves of other times."[2] At other times, she rebels at the necessity of playing a role of someone else's designing: "I must be careful not to speak aloud, though, for if I do Marvin will look at Doris and Doris will look meaningfully back at Marvin, and one of them will say, 'Mother's having one of her days.' Let them talk. What do I care now what people say? I cared too long."[2]

Hagar's occasionally disconcerting behavior, her family's uneasiness with her physical slowness, and the extra demands they feel all of this puts on them lead Marvin and Doris to urge Hagar to move into a nursing home even before they learn of her illness. In fact, they take

Hagar to Dr. Corby in the hope that he will provide the needed medical ammunition to defeat Hagar's protestations against moving to Silverthreads. Against such organized opposition, Hagar feels helpless. "Can they force me?" she wonders. "I glance from one to the other, and see they are united against me. Their faces are set, unyielding. I am no longer certain of my rights."[2] Caught in such a position, Hagar begins to see herself in her family's eyes, to feel ashamed of her body and its functions. When Doris must help her dress for bed, Hagar is annoyed at having to hold on to her daughter-in-law's hand for balance and have her corsets removed by someone else, but she feels angry humiliation at having to appear naked before Doris, to have, as Hagar puts it, "her see my blue-veined swollen flesh and the hairy triangle that still proclaims with lunatic insistence a non-existent womanhood."[2] There are echoes in this passage of the "hag" that Roethke's old woman sees in others' eyes when they look at her. Hagar is seeeing her body as Doris sees it because Doris is forcing her unsympathetic presence and help upon Hagar.

Hagar responds to the pressure by running away. She crosses town and hides in an abandoned building. She is uncomfortable because of the physical demands such an adventure makes on her body, but it is eventually her illness that debilitates her to the point where she must abandon her escape and enter a hospital. Upon finally hearing the name of her disease, Hagar responds with decision and firmness. She comments, "Odd. Only now do I see that what's going to happen can't be delayed indefinitely."[2]

Hagar spends the remaining weeks of her life learning about real, painful illness. Her experience with illness is different from her experience with aging, even in terms of her relationship with her family. Hagar can see Marvin's fears and needs and is no longer threatened by them. Marvin now seeks her understanding rather than her compliance. She is once more in control of her life, even though she is in the face of death. The kind of thinking that led Marvin and Doris to define old age as unhealthy not only denied Hagar the means of living her old age with freedom and dignity, but also filled Hagar's mind with confusion and mistrust.

The problems Barbara Macdonald has encountered because of others' definitions of her supposed incompetence have also affected her self-confidence and communication with others. In one essay Macdonald describes her participation in a rally to "take back the night," a march focusing attention on women's vulnerability to rape and other violence. A physically active woman herself, walking through the Boston streets with a large group of women, even at night and in the rain, was no cause for concern to Macdonald. She, therefore, did not immediately perceive the problem when her younger companion was drawn aside by one of march's monitors. Macdonald overheard them at first with some confusion. Then:

> The monitor was at first evasive and then chose her words with care, "if you think you can't keep up, you should go to the head of the march." Gradually, I took it in, like a series of blows, what the situation was, that the monitor had thought that, because my hair is grey, because I am sixty-five and because I look sixty-five, I might, if slower, leave a gap between the ranks . . . and that she could not say this to me. I stepped directly in front of the monitor for eye to eye contact to force her to talk *to me* instead of *about me* . . . [1] (emphasis mine).

Macdonald's immediate reaction to the monitor was one of rage. Then the rage was replaced by a more debilitating emotion, a feeling of caution, bread out of "the dread of being told [she] did not belong there."[1] She began to doubt the sympathy and support of the women with whom she was marching. She reports that these doubts led her to see herself from the eyes of these other women, and to see herself unfavorably: "My short stature, my grey hair, my wrinkled face—I wondered how sixty-five years looked to them. And finally I looked at the four other women who were to walk beside me. I wondered how they felt about being with me."[1]

As Macdonald saw her small, wrinkled self through others' eyes she began to doubt her very worth as a person; she wondered if people even wanted to be with her. Although she felt her "own muscles and [kept] reassuring [herself], 'Barbara, you don't feel weak to me,' " she still had trouble believing what she knew about her own strength, and thus, about her health. "I wondered how to feel proud and strong," she writes, "when women around me were telling me I was weak."[1] She finally concluded that the issue was not necessarily one of suspected weakness, but rather that the monitor feared that Macdonald was unable to judge whether or not she was capable of the task. "She believes," said Macdonald, "that a sixty-five-year-old woman lacks judgment about what she can do. She thought I did not perceive the situation and that I did not know what I was doing."[1]

Either reason for the monitor's behavior echoes Laurence's portrayal of Hagar's situation. Macdonald was being defined as mentally as well as physically incompetent solely because of her age. She was avoided because her presence made other, younger people uncomfortable and perhaps inconvenienced. Their false judgments caused Macdonald to doubt her own physical abilities and even her personal worth.

Macdonald finished the march, but the enthusiasm had died and she felt instead a "rage that had no place to go." Her rage, however, soon gave way to fear. She writes, "As I took in the inevitability of my becoming less and less able to protect myself, all I could feel was a kind of hopelessness and panic."[1] Her body became the enemy, a failing security system leaving her vulnerable to the actions and judgments of others, much as Hagar's old body became a personal rebuke to her when forced to be helped to bed by her daughter-in-law.

Ultimately, however, Macdonald was not defeated by fears and doubts created by other people's visions of her. She realized that, as a strong-willed woman and a lesbian, she had always had to fight against the views of stronger majorities. "This was no new experience that was going to come with aging,"[1] she writes. Like Hagar, Macdonald needed to understand the situation in which she found herself. Hagar needed to know which pains were to be expected of aging and which were to be recognized as invading illness. Macdonald needed to distinguish which physical dangers were inescapable with aging and which emotional threats she could grid herself to fight, as she had been fighting all her life.

When Hagar discovered her actual illness, she could bring all her ageless strength to meet the situation. Likewise, Macdonald, in discovering a new threat to her emotional well-being, could now summon her entire self to meet the new challenge. She proclaims:

> Thus I healed myself and could feel whole again, connected to my aging body, wanting to live my life out in partnership with it, without feelings of humiliation because of its difference, and without the fear that I would so want to disclaim it that I would fail to protect it.[1]

Macdonald is truly, as was Roethke's old woman, singing "from the folds of [her] skin." Her triumph, as it is for all three of the women considered here, is not over her old age, but over what other people would make of old age. It is a triumph that requires growing aware of and happy with the changes that come naturally with aging. It is a triumph because one must achieve this awareness and acceptance in the face of countless other people who view old age as weakness, illness, and incompetence.

The three women presented here are able to triumph because they are able, first, to dissociate their minds from their bodies long enough to recognize their physical changes, and then reintegrate their physical and emotional beings into one person who is not merely comfortable, but also pleased with her aged identity. The people in these literary works who responded negatively to Hagar Shipley, Barbara

Macdonald, and to some extent, Roethke's old woman did not go through this process of dissociation and reintegration in terms of their perceptions of these aging women. The pictures of aging presented by these three writers suggest that definitions of health must be dynamic rather than static. When definitions of health do not change with changing circumstances, emotional tensions can arise within the aging person, within the people around her or him, and between both parties. The aging person becomes isolated at a time when social, physical, and emotional support is most needed. Mutual acceptance and appreciation of aging is necessary in order to aid the transition of an aging woman or man to a new—though not necessarily unhealthy—stage of life.

Notes and References

[1]Cynthia Rich, Cynthia's introduction, in *Look Me in the Eye: Old Women, Aging and Ageism,* (Barbara Macdonald and Cynthia Rich, ed.), Spinsters, Ink., 803 DeHaro St., San Francisco, California 94107, p. 10 (quoted with permission from the authors).

[2]Margaret Laurence (1964) *The Stone Angel* Alfred A. Knopf, New York.

[3]Theodore Roethke (1975) I'm here, from Meditations of an old woman, in *The Collected Poems of Theodore Roethke* Doubleday, New York.

Index